Cursor: The Complete Guide

The AI Code Editor — From Tab Completion to Agent-Driven Refactors

Ken Tannenbaum

2026

Finnoybu Press

Cursor: The AI Code Editor

2026 Edition

Cursor — From Tab Completion to Agent-Driven Refactors

Published by Finnoybu Press

For permissions and inquiries: press@finnoybu.org

First Edition: 2026

About the Author

Ken Tannenbaum is a technologist, entrepreneur, and AI governance researcher. He is the founder of Finnoybu Press and the founder/steward of the AEGIS Initiative, an open framework for AI safety and accountability. Learn more at aegis-initiative.com.

When he's not building governance standards for artificial intelligence, he's helping everyday people understand and use it.

Dedication

For every developer who has stared at a blinking cursor at 11 p.m., trying to remember whether to fight the AI, ignore it, or let it drive — and for the ones who figured out that the answer changes by the task.

This book is about figuring it out faster.

The AI for Everyone Series

This book is part of the **AI for Everyone** series by Finnoybu Press, designed to help real people use AI tools effectively. Visit **press.finnoybu.org** for the latest titles and updates.

Collection One: Product Deep Dives

- *ChatGPT: The Definitive Guide (2026 Edition)*
- *ChatGPT: The Power User Guide (2026 Edition)*
- *Claude: The Definitive Guide (2026 Edition)*
- *Claude Code: The Power User Guide (2026 Edition)*
- *Gemini: The Complete Guide (2026 Edition)*
- *Gemini: The Advanced Guide (2026 Edition)*
- *Copilot: The Workplace AI Guide (2026 Edition)*
- *Copilot: The Advanced Workplace Guide (2026 Edition)*
- *Perplexity: The Research Playbook (2026 Edition)*
- *Perplexity: The Advanced Research Playbook (2026 Edition)*

Collection Two: Product Guides

- *NotebookLM: From Sources to Audio Overviews (2026 Edition)*
- *DeepSeek: Code, Math, and Reasoning (2026 Edition)*
- *Cursor: The AI Code Editor (2026 Edition)*
- *OpenClaw: Agentic AI for Everyday Work (2026 Edition)*

Collection Three: Cross-Platform Skills

- *AI Systems Playbook (2026 Edition)*
- *AI Prompt Engineering (2026 Edition)*
- *AI for Developers (2026 Edition)*
- *AI for Business (2026 Edition)*
- *AI for Students (2026 Edition)*
- *AI Governance for Practitioners (2026 Edition)*
- *Securing Agentic AI (2026 Edition)*

Table of Contents

Introduction

I wrote this book because Cursor is the most-used new developer tool of the last two years and almost nobody using it is using it well.

That's not a knock on the people using it. Cursor moves fast — three named modes, a Tab model, a Cmd+K inline editor, four frontier models in the picker, Project Rules, codebase indexing, MCP servers, Background Agents, BugBot, Multitask, and a credit-pool pricing model that can drain $20 in a long Tuesday. The product changes every few weeks. Community advice on Reddit, Twitter, and YouTube is half-current at best, and the official docs cover features one at a time without telling you which features matter for what kind of work.

The result: most developers using Cursor have settled into a partial workflow. They use Tab and ignore Agent. Or they use Agent for everything, including work it isn't good at, and burn credits while wondering why the output gets worse the longer the session runs. Or they read about Project Rules, didn't quite get them set up, never came back, and now their team's `.cursor/rules/` folder is empty.

This book is about getting past partial. It covers every mode, every model, every advanced feature, and — more importantly — the workflow discipline that separates effective Cursor users from the ones fighting it. The 30%-to-80% acceptance-rate jump that good rules and good prompts unlock isn't a feature you turn on. It's a habit. This book is about that habit.

Who This Book Is For

This book has a wider audience than its developer-focused subtitle implies:

- **Solo developers** building side projects or working as contractors who need Cursor to feel like leverage, not a credit drain
- **Engineers on small teams** who want shared rules and a workflow that scales beyond one person
- **Tech leads and engineering managers** evaluating Cursor against Claude Code, Windsurf, and other AI editors for their teams
- **CTOs and engineering directors** thinking about Cursor at the org level — Teams, Enterprise, SCIM, audit logs, compliance posture

- **Anyone who has installed Cursor** and feels like they're using maybe 30% of what it can do
- **Anyone curious** about where AI-assisted development is heading and what an "AI-native IDE" means in practice

You'll get the most from this book if you've written code recently — the walkthroughs assume you can read TypeScript, Python, or Go without translation. But the strategy chapters (Parts I, V, VII, VIII) are accessible to engineering leaders who don't write code daily but need to make decisions about it.

How to Read This Book

The book is organized in nine parts:

Part I: Getting Started (Chapters 1–5) introduces what Cursor is, how it compares to the alternatives, which plan to choose, how to install and configure it, and how to make your first edits with AI.

Part II: The Three Modes (Chapters 6–9) covers Ask, Manual Edit (formerly Composer), and Agent — what each is for, when to use which, and how to switch between them mid-task.

Part III: Tab and Inline Edits (Chapters 10–12) is the inline-completion core — Tab autocomplete, Cmd+K, and the diff-review habit that protects you from bad accepts.

Part IV: Working at Scale (Chapters 13–17) is where Cursor becomes leverage instead of a toy — codebase indexing, Project Rules, `.cursorignore`, Background Agents, and `/multitask`.

Part V: Choosing and Using Models (Chapters 18–21) covers the four-model picker, credits and usage-based pricing, Bring Your Own API, and Privacy mode.

Part VI: The Cursor Ecosystem (Chapters 22–25) covers MCP, BugBot, hooks and slash commands and Skills, and the terminal/debugger.

Part VII: Cursor for Your World (Chapters 26–28) applies Cursor to specific contexts — solo, team, and enterprise — with realistic walkthroughs for each.

Part VIII: Safety, Ethics, and Limitations (Chapters 29–32) is the section other coverage skips: what Cursor gets wrong, code review and trust, security and compliance, and the ethics of AI-assisted development.

Part IX: Reference (Appendices A–F) gives you keyboard shortcuts, a 40+ entry starter rules library, troubleshooting, glossary, the 2026 timeline, and resources.

Read straight through or jump to whatever fits your work right now. Solo developers should read Parts I, II, III, and Chapter 26, then reference the rest. Team leads should read Parts I, IV, V, VII, and VIII. Engineering directors and security teams should focus on Parts V, VII, and VIII. For the full picture, read straight through — the chapters build on each other in roughly the order a developer would adopt the features.

A Note About Accuracy

Cursor changes faster than almost any tool this book series has covered. The Cursor team ships new features, model picker updates, pricing changes, and capability expansions on a cadence that has occasionally embarrassed even the most attentive coverage. This book is accurate as of early 2026, but specific details — exact pricing, current model versions, feature availability, the UI of a given panel — may have changed by the time you read this.

The principles, techniques, and workflow patterns have a longer shelf life than the numbers. How to scope a task before invoking Agent, how to write a Project Rule that shifts model behavior, how to read a proposed diff critically, how to choose between Ask and Manual and Agent — these skills carry across versions and even across other AI editors.

When something looks different from what's described here, check the official Cursor documentation at docs.cursor.com.

A Note About Cost

Cursor's pricing changed materially in 2025 with the shift from request quotas to credit pools. A heavy day on Claude Opus 4.7 in Agent mode can drain a Pro subscription's credit pool in three days. This book takes cost seriously — Chapter

19 is dedicated to it, and most chapters note where a feature is cheap, expensive, or surprisingly expensive.

If you're going to use Cursor seriously, understand its economics. A team of ten engineers on Pro plans can quietly burn $2,000+ a month, and the wrong defaults can double that without anyone noticing. The book's recommendations on model selection, BYOK, and rules-driven workflows are partly about quality and partly about not lighting money on fire.

Let's Get Started

The 2026 Cursor is not the Cursor you tried in 2024. The product has graduated from "VS Code with a chat sidebar" to a full agent execution runtime that happens to also be an editor. The workflow that gets the most out of it differs from the one that got results last year, and the workflow that will work in 2027 will probably differ again.

What's durable is the discipline. Scope before you prompt. Read every diff. Write rules instead of repeating yourself. Match the mode to the task. Treat agents as fast junior developers who need a clear ticket and careful review.

Turn the page. Let's start with what Cursor is.

Part I: Getting Started

"The tool does not make the craftsman, but the wrong tool will quietly ruin the work."

— A working principle

Chapter 1: What Is Cursor?

In the spring of 2023, four MIT graduates released the first public version of Cursor — a fork of Visual Studio Code with a few AI features bolted on. It was a sidebar with chat, an inline edit shortcut, and a tab-completion model that occasionally guessed what you wanted. Perhaps a dozen comparable products launched in the same six-month window. Most have since been forgotten.

Cursor stuck. By mid-2024, it was the AI editor a noticeable share of working developers had switched to. By late 2025, it had crossed a million paid subscribers, signed enterprise contracts at firms whose internal tooling teams normally took years to approve a new editor, and shipped Background Agents — cloud-hosted agents that worked autonomously for tens of minutes at a time on real codebases. By early 2026, Cursor had graduated BugBot from beta to a general-availability automated reviewer-and-fixer, integrated DeepSeek V4 Pro into the model picker, and shipped 3.0 with `/multitask` parallel async subagents.

If you're reading this book, you've probably encountered Cursor through one of three doors. Maybe you tried it on a side project and noticed Tab autocomplete was uncannily accurate compared to what you'd seen elsewhere. Maybe a coworker mentioned that Cursor's Agent mode finished a refactor they didn't want to do themselves. Maybe your engineering org evaluated half a dozen AI editors, Cursor came out ahead, and now you're trying to figure out how to use it well.

Whatever brought you here, this chapter gives you a clear picture of what Cursor is, how it differs from the tools you've been using, and what makes it consequential beyond the hype cycle.

What Cursor Is

Cursor is an AI-native code editor — a desktop application, distributed as a fork of Visual Studio Code, that rebuilt the editor's interaction surface around AI assistance. It's built and operated by Anysphere, the New York–based company the four MIT founders started after their initial release.

The "AI-native" framing is doing real work in that sentence, and it's worth unpacking.

A traditional editor is a tool for typing characters into files. AI features, when present, layer on as extensions or sidebar panels running alongside the editor's core. This is what GitHub Copilot does inside vanilla VS Code: a separate process suggests completions through an extension API, while the editor itself remains a character-typing tool that doesn't know AI exists. The AI is a guest in the editor's house.

Cursor treats AI as the primary interaction surface. The editor's chat panel isn't a sidebar feature — it's a first-class workspace that can read your codebase, propose edits across multiple files, run terminal commands, fetch web pages, and execute multi-step plans. Inline editing through Cmd+K is built into the editor's core, not an extension. The Tab autocomplete model is owned by Cursor and tuned to its workflow. Project Rules in `.cursor/rules/` are read by every model invocation by default. The AI isn't a guest; it's the host, and you're a developer who's been given keys to the building.

What Cursor ships comes in four forms:

The desktop editor. A 200 MB cross-platform app for macOS, Windows, and Linux, distributed at cursor.com. It's a downstream fork of VS Code, which means almost every VS Code extension works in Cursor and the keybindings, themes, and basic UX feel familiar to anyone coming from VS Code. Think of it as "VS Code plus AI primitives" without losing the rest.

A subscription service with a free Hobby tier and paid Pro, Pro+, Ultra, Teams, and Enterprise plans. The subscription gates access to the frontier models in the picker (Claude Opus 4.7, GPT-5.4, DeepSeek V4 Pro, Gemini 3.1 Pro), the credit pool used by usage-based billing, and team and enterprise features like shared rules, SCIM, and audit logs. Chapter 3 covers the plans in detail.

Background Agents. Cloud-hosted agents that run on Cursor's infrastructure, work for ten to thirty minutes on a task, and report back when done. Kick one off and walk away to lunch. Background Agents are the feature that pushed Cursor from "smart editor" into "agent runtime" territory. We cover them in Chapter 16.

A growing ecosystem of MCP servers, Project Rules templates, custom slash commands, hooks, and Skills — the parts of Cursor that you and your team configure and share. The ecosystem turns Cursor from a personal tool into a team-wide workflow standard, and Part VI covers it end-to-end.

The same underlying editor powers all four. When you write a Project Rule, Background Agent reads it. When you install an MCP server, every model invocation can use its tools. The integration is the point.

What Cursor Is Not

Understanding what Cursor *isn't* matters for using it well. A few common misconceptions worth clearing up:

Cursor is not a chatbot in a sidebar. It looks like one if you only use the chat panel, but the chat panel is the smallest part of the product. Most of Cursor's value comes from Tab, Cmd+K, Project Rules, Agent mode, Background Agents, and MCP — none of which live in the sidebar. If you've been using Cursor for a year and only ever interacted with the chat panel, you've been using maybe 20% of what's there.

Cursor is not "Anthropic's editor" or "OpenAI's editor." Cursor is built by Anysphere and is model-agnostic in the picker. You can route the same workflow through Claude, GPT, DeepSeek, or Gemini, and the right choice changes by task. Chapter 18 covers the picker in detail. The model loyalty most engineers carry from chat experiences ("I always use Claude" or "GPT is the best") rarely survives a week of comparing models on the same Cursor task.

Cursor is not a replacement for thinking. Agent mode is good. It's not a substitute for understanding what you're asking it to do. The teams getting the most leverage out of Cursor have the highest standards for code review, test discipline, and architectural clarity — not the ones trying to delegate those skills to the model. The book returns to this in Chapter 30.

Cursor is not free for serious use. Hobby tier exists and works for trivial tasks. Anyone using Cursor in their day job moves to Pro or higher within a week or two. Chapter 3 walks through the plan tradeoffs, and Chapter 19 walks through credits and usage-based pricing — including the "$20 in three days" trap that catches new Pro subscribers who haven't tuned their workflow yet.

Cursor is not the only good answer. Claude Code is excellent for terminal-native workflows. Windsurf is a real competitor some teams prefer. Vanilla VS Code with Copilot is sufficient for some contexts. This book doesn't pretend Cursor is universally the right tool — Chapter 2 gives you a decision framework for when it isn't.

A Brief History

Cursor's trajectory ranks among the fastest in developer-tooling history, and the rough timeline helps explain where the product is now.

The 2023 Launch

Cursor 0.1 shipped in March 2023. The initial product was a VS Code fork with three AI primitives: chat in a sidebar, Cmd+K for inline edits, and Tab autocomplete. Early traction was modest — it was one of half a dozen "AI editor" products competing for attention in the wake of ChatGPT's launch. The founders were on YC W23 under the company name Anysphere.

What set the early Cursor apart was a willingness to ship aggressively and ignore the "we should be a VS Code extension" path most AI tooling startups took. By forking VS Code, the team could change anything — the rendering of inline diffs, how the editor handled multi-file proposals, the keybindings, the data model — without depending on Microsoft's extension API. That tradeoff had costs (every VS Code update needed re-merging) but gave Cursor design latitude extension-based competitors couldn't match.

The Composer Era (Late 2024)

The first feature that made Cursor distinctive arrived in late 2024 with Composer. Composer let you describe a multi-file change in natural language and watch a model propose a coordinated set of edits across files, with a unified diff-review interface. This was the moment Cursor stopped being "VS Code with chat" and became "VS Code with multi-file AI editing." Composer would later be renamed Manual Edit mode and joined by Agent mode, but the core capability — coordinated multi-file proposals you could review and accept — became the spine of Cursor's editor experience.

Agent Mode and Background Agents (Mid-2025)

In the first half of 2025, Cursor split its editor experience into the three-mode model: Ask (read-only exploration), Manual Edit (model proposes diffs you review), and Agent (autonomous execution). Agent mode could read files, edit files, run terminal commands, and fetch web pages on its own — a step beyond what Composer/Manual could do. Around the same time, Background Agents shipped: cloud-hosted Agent runs that worked for tens of minutes without your editor open.

These two features marked a substantial shift. Cursor moved from "smart editor with chat" to "editor with full agent execution" and "agent execution that runs in the cloud while you do something else." The implications for how teams worked through a backlog were significant — tickets could be assigned to agents the way they'd been assigned to junior engineers, with similar review patterns.

Pricing Model Transition (Mid-2025)

In June 2025, Cursor changed its pricing from request quotas to credit pools. Pro went from "500 fast requests per month plus unlimited slow" to "$20 of usage credits per month at usage-based prices." The shift was contentious — some heavy users paid substantially more, and a flurry of refund discussions and pricing adjustments followed. The credit-pool model is the current state and reflects the underlying economics of a workflow where a single Agent task can consume substantial model time. Chapter 19 covers the implications.

BugBot Beta (Late 2025)

In late 2025, Cursor released BugBot in beta — a GitHub-integrated automated PR reviewer that read pull requests and posted review comments. BugBot was a separate add-on with its own pricing tier. Useful but limited; in beta, it identified issues but didn't fix them.

BugBot Fixer and Cursor 3.0 (Early 2026)

Two things in the first quarter of 2026 materially changed the product. First, BugBot graduated from reviewer to fixer in February, with reported 80%-plus resolution rates on issues it identified — meaning BugBot now opens PRs of its own to fix problems it finds. Second, Cursor 3.0 shipped with `/multitask`, a slash command that spawns parallel async subagents in isolated worktrees, lets them work independently, and lets you pull the winner into the foreground. Multitask lets you "try three approaches at once" without managing the orchestration yourself. We cover both in detail later in the book.

Around the same time, the model picker added DeepSeek V4 Pro as a frontier reasoning option at roughly one-tenth the cost of Claude Opus, expanded MCP one-click installs from a curated catalog, and shipped a series of indexing performance improvements aimed at monorepos. The 2026 Cursor differs meaningfully from the 2025 Cursor, and the gap will keep widening.

Where Cursor Fits in the AI-Coding Landscape

Cursor is one of several major options for AI-assisted development in 2026. Understanding where it fits helps you choose the right tool — and decide whether Cursor specifically is the right tool for your work:

- **GitHub Copilot in VS Code** — The lowest-friction option, especially for organizations already standardized on VS Code. Strong Tab autocomplete, weaker on multi-file refactors and agent workflows. Covered in *Copilot: The Workplace AI Guide* and *Copilot: The Advanced Workplace Guide*.
- **Claude Code (terminal)** — Anthropic's terminal-native agent. No editor — it's a CLI that operates on your filesystem. Strong on long, autonomous tasks; weaker on the inline-editing experience. Covered in *Claude Code: The Power User Guide*.
- **Windsurf** — The closest IDE rival to Cursor, also a VS Code fork with AI-native features. The product comparison is close enough that many teams pilot both. Chapter 2 covers the differences.
- **Cline, Continue.dev, Aider, Void AI** — The open-source and BYOK alternatives. Lower-cost or zero-cost paths if you're willing to invest setup time. Most lack the polished editor experience and the team features.
- **Cursor** — The leader in the AI-native-IDE category as of early 2026. Most polished agent and rules tooling, broadest model picker, deepest team and enterprise features.

A common pattern in production engineering organizations is **multi-tool routing**: Cursor as the default editor, Claude Code in the terminal for long autonomous tasks where you don't want an editor open, and Copilot retained for engineers who haven't switched. This isn't a problem — different tools fit different work. The book covers Cursor's strengths and limits honestly so you know which work fits Cursor and which doesn't.

The head-to-head comparison comes in Chapter 2.

What You'll Learn in This Book

This book is organized to take you from a curious user who's installed Cursor to a confident user who can work effectively across modes, models, scales, and team contexts:

- **Part I (Chapters 1–5)** gets you oriented: what Cursor is, the alternatives, the plans, installation and configuration, and your first edits.
- **Part II (Chapters 6–9)** covers the three modes — Ask, Manual, Agent — and how to choose between them.
- **Part III (Chapters 10–12)** is the inline-editing core: Tab, Cmd+K, and reading diffs critically.
- **Part IV (Chapters 13–17)** is where Cursor scales — codebase indexing, Project Rules, ignore files, Background Agents, and Multitask.
- **Part V (Chapters 18–21)** covers models, credits, BYOK, and Privacy mode.
- **Part VI (Chapters 22–25)** covers the ecosystem — MCP, BugBot, hooks/commands/Skills, and the terminal/debugger.
- **Part VII (Chapters 26–28)** applies Cursor to solo, team, and enterprise contexts.
- **Part VIII (Chapters 29–32)** covers the limits — what Cursor gets wrong, code review and trust, security and compliance, and ethics.
- **Part IX (Appendices A–F)** gives you reference material for ongoing use, including a 40-plus starter library of `.cursor/rules/` files.

Every chapter is designed to be useful on its own. Solo developers should skim Part I, work through Parts II and III carefully, and reference the rest. Engineering leads should focus on Parts I, IV, V, and VII. For the full picture, read straight through.

Let's get started. Chapter 2 covers how Cursor stacks up against VS Code with Copilot, Claude Code, Windsurf, and the open-source alternatives — so you can decide whether it's the right tool for your work before spending the rest of the book learning to use it well.

Chapter 2: Cursor vs. VS Code, Copilot, and the Alternatives

Before you spend the rest of this book learning to use Cursor well, you should be confident that Cursor is the right tool for your work. It usually is, for the kind of developer who picks up a book like this — but not always. This chapter walks through the real alternatives, what each is good at, and the decision framework that should drive the choice.

The chapter is honest about Cursor's limits. If your work is better served by Claude Code, Windsurf, or vanilla VS Code with Copilot, the book will tell you that here, and you can save the price of the rest of it.

The Comparison That Matters

The category isn't "AI assistant." Every modern editor has an AI assistant. The category is "where does the AI live in your workflow," and there are four meaningfully different answers in 2026:

1. **AI as an extension** — VS Code with Copilot, JetBrains with Copilot or AI Assistant, Neovim with various plugins. The editor is unchanged; AI is a guest.
2. **AI as the primary surface in an editor** — Cursor, Windsurf. The editor itself was rebuilt around AI primitives; the AI is the host.
3. **AI as a terminal-native agent** — Claude Code, Aider. No editor; you talk to the AI in your terminal and it operates on your filesystem.
4. **AI as a per-session web app** — chat.openai.com, claude.ai, copy-paste workflows. Useful for quick questions; not a full development workflow.

Cursor sits firmly in category 2. When you choose Cursor, the decision is whether the value of having AI as the primary surface beats the friction of switching editors and the cost of the subscription. For most working developers in 2026, it does. For some, it doesn't. The rest of this chapter helps you figure out which group you're in.

Cursor vs. Vanilla VS Code with Copilot

This is the comparison most readers care about, since most developers considering Cursor are coming from VS Code.

The shared ground is large. Cursor is a VS Code fork, so the editor surface — file tree, command palette, terminal panel, Git integration, debugger, extensions marketplace — is essentially identical. Almost every VS Code extension installs and works in Cursor. Themes, keybindings, and settings transfer with a one-click import on first launch. If you're switching from VS Code to Cursor, you're not learning a new editor; you're getting a familiar editor with new primitives layered in.

The differences are in the AI features:

Tab autocomplete. Both products have it. Copilot's Tab is good. Cursor's Tab is better in most contexts I've tested — it has a longer effective context window (it sees more of your codebase when proposing a completion), handles multi-line completions that span structural boundaries better, and its "smart rewrites" can propose edits to lines you didn't have selected. The gap is smaller than it was a year ago — Copilot has been catching up — but Cursor's Tab still feels noticeably more codebase-aware in monorepo contexts.

Inline edits. Cmd+K in Cursor is a first-class feature with a tuned UX: select code, hit the shortcut, type your prompt, and get a proposed diff inline you can accept or reject. Copilot's inline edit (Copilot Edits, accessed through a chat sidebar workflow) is less integrated and adds more friction. Cursor feels closer to "talking to your editor"; Copilot feels closer to "asking the chat panel to make changes."

Multi-file edits. Cursor's Manual Edit and Agent modes can propose coordinated changes across many files in a single review session. Copilot Edits has comparable multi-file support but the diff-review experience is less polished, and the model's ability to coordinate across more than five or ten files at once is weaker in my experience.

Agent mode. Cursor's Agent mode is a substantial feature gap. Copilot has Agent Mode in VS Code, and it's improving, but the workflow tooling around it — the rules system, codebase indexing, how Agent integrates with your terminal and external tools — is materially less mature. If you want to assign tickets to an agent and let it work autonomously while you do something else, Cursor's product is meaningfully ahead.

Project Rules. Cursor's `.cursor/rules/` system, covered in detail in Chapter 14, is a first-class feature with substantial leverage. Copilot has a comparable instructions system that's useful, but Cursor's tooling around glob-scoped rules, rule references, and the rules-as-team-asset pattern is more developed.

Background Agents. Cursor has Background Agents — cloud-hosted agents that work for tens of minutes without your editor open. Copilot has GitHub Copilot Workspace and the Copilot coding agent, which serve a similar use case, and the gap here has narrowed considerably as GitHub has invested in the agent space. If you're already paying for GitHub Enterprise, Copilot's coding agent integration with GitHub Issues and Actions is worth taking seriously.

Pricing. Copilot starts at $10/month for individuals. Cursor Pro is $20/month, Pro+ is $60/month, Ultra is $200/month. For a heavy daily user, Cursor's costs can grow into the hundreds per month if you're using premium models in Agent mode without rules-driven workflow discipline. Copilot's pricing is more predictable; Cursor's scales with your work, in both directions.

The deciding question: Are you using AI for the occasional autocomplete and the rare refactor, or as a substantial part of how you work? If the former, vanilla VS Code with Copilot is fine and probably cheaper. If the latter, Cursor's investment in agent workflows, rules, and multi-file editing will compound over months of use in ways Copilot won't match in the same timeframe.

Cursor vs. Claude Code

Claude Code is Anthropic's terminal-native agent. No editor — you run a CLI in your terminal, it reads your filesystem, edits files, runs commands, and talks to you through the terminal interface. We covered Claude Code in detail in *Claude Code: The Power User Guide*; this section is about when Cursor is the right choice and when Claude Code is.

The cases for Claude Code over Cursor:

- **Long autonomous tasks without an editor open.** Claude Code shines at "scope this ticket, plan the work, implement it, run tests, iterate until passing" workflows where you want to assign work and walk away.

Cursor's Background Agent serves a similar role, but Claude Code's CLI workflow is more streamlined for fully headless operation.

- **Server and CLI environments.** If you're working on a remote server, Claude Code over SSH is materially easier than running Cursor with Remote-SSH. The terminal-native model is the right model for that context.
- **Agent SDK use cases.** If you're building automation around the agent — wrapping it in your own workflows, calling it from CI, treating it as an SDK rather than a tool you sit in front of — Claude Code's CLI shape fits better than Cursor's editor-first design.
- **Lower friction for short tasks.** "Open a terminal in this folder and ask the agent to fix the failing test" is faster in Claude Code than Cursor for some workflows, because there's no editor state to manage.

The cases for Cursor over Claude Code:

- **Inline editing as part of the workflow.** Tab autocomplete, Cmd+K inline generation, and Manual Edit's diff-review interface have no equivalent in Claude Code. If your work is editor-shaped and not agent-shaped, Cursor is the right tool.
- **Multi-model picker.** Claude Code is single-vendor; Cursor's picker includes Claude, GPT, DeepSeek, and Gemini. For tasks where you want to try the same prompt across models or where a non-Claude model is the better fit, Cursor's flexibility is real.
- **Visual diff review and selection-based prompts.** Reading a proposed multi-file diff in a tree view and clicking through hunks one at a time is a meaningfully different experience from reviewing a CLI agent's edits. For careful review of large changes, Cursor's UX wins.
- **Team rules and shared configuration.** Cursor's rules-and-skills tooling at the team level (covered in Part VII) is more developed for shared editor workflows than Claude Code's equivalents.

The honest answer: many engineers use both. Cursor for editor-shaped work, Claude Code for terminal-shaped work. They're not fungible.

Cursor vs. Windsurf

Windsurf (formerly Codeium) is the closest direct competitor to Cursor: also a VS Code fork, also AI-native, also a chat-plus-edit-plus-agent product. The two have leapfrogged each other on features through 2025 and into 2026, and the choice between them is closer than the choice between Cursor and any other alternative.

What Windsurf does well:

- **Cascade.** Windsurf's flagship agent feature is well-regarded, with a workflow some users prefer to Cursor's Agent. The "agent that watches what you're doing and proactively offers help" model is more developed in Windsurf.
- **Predictable pricing.** Windsurf's pricing has been more stable than Cursor's, with less of the credit-pool volatility that frustrated some Cursor users in mid-2025.
- **Enterprise positioning.** Windsurf has invested heavily in enterprise sales and won significant accounts, particularly in regulated industries.

What Cursor does better:

- **Model picker breadth and polish.** Cursor's picker, the BYOK story, and per-model behavior tuning are more mature.
- **Rules system.** Cursor's `.cursor/rules/` tooling outpaces Windsurf's equivalent.
- **Agent ecosystem.** Background Agents, BugBot, Multitask, and the MCP catalog are richer in Cursor.
- **Community gravitational pull.** More tutorials, more shared rules libraries, more YouTube content, a larger user community to ask for help.

If you're starting from scratch and either tool would fit, Cursor is the safer choice based on ecosystem momentum. If you're already on Windsurf and it's working, switching costs may not be worth it. Many teams pilot both for a week and pick whichever the team gravitates to.

Cursor vs. the Open-Source Alternatives

A handful of open-source AI editors compete in roughly the same space:

- **Cline** (formerly Claude Dev) — a VS Code extension that does Agent-mode-style autonomous coding. Free, BYOK (you bring your own API key), runs in vanilla VS Code. Good for users who already have an Anthropic or OpenAI key and want agent capabilities without paying Cursor's subscription.
- **Continue.dev** — an open-source AI assistant for VS Code and JetBrains. Highly customizable, BYOK, supports many model providers. Less polished than Cursor; more flexible.
- **Aider** — a CLI agent in the same space as Claude Code, with a longer history. Open source, BYOK, capable for terminal-native workflows.
- **Void AI** — a more recent open-source Cursor-alternative gaining attention. Earlier in maturity; one to watch.

The case for any of these over Cursor: cost. If you have your own API key and don't want a subscription, you can get a meaningful share of Cursor's value through Cline or Continue.dev for the price of your API usage. The case against: setup complexity, lower polish, smaller ecosystem, less mature team and enterprise features, less curated MCP integrations.

The right call depends on your engineering preferences. If you enjoy configuring tools and want fine-grained control, the open-source alternatives are real options. If you want an editor that works out of the box and you're willing to pay for the polish, Cursor wins.

A Decision Matrix

A rough framework for choosing among the alternatives, organized by what kind of developer you are:

Solo developer, side projects only, occasional AI use. Vanilla VS Code with Copilot at $10/month. Cursor's Hobby tier also works. Don't pay for a Pro subscription you won't use.

Solo developer, paid client work, AI-heavy workflow. Cursor Pro at $20/month, with discipline around model selection. The credit pool covers a productive month if you're not running Opus on every request.

Solo developer, very heavy AI use, willing to manage tools. Cline or Continue.dev with your own Anthropic or OpenAI key, paying per token instead of per subscription. Cheaper at scale, more setup work.

Small team (2–10 engineers), shared editor workflow. Cursor Teams at $40/user/month. The shared rules and centralized billing pay for themselves quickly compared to managing individual subscriptions.

Small team, terminal-native culture. Claude Code with Anthropic billing, possibly with a few Cursor seats for engineers who prefer an editor.

Large team / enterprise, regulated environment. Cursor Enterprise or Windsurf Enterprise after careful evaluation. Both are credible. The choice usually comes down to which one your security team is more comfortable with after threat modeling — Chapter 21 and Chapter 31 cover what to look at.

You're building automation around an agent. Claude Code or the Anthropic Agent SDK directly. Cursor's editor-first model isn't built for headless automation.

You're a long-time vim or Emacs user and the editor itself is the workflow. Stay where you are with one of the AI plugins for your editor of choice. The cost of switching editors is real and the returns from Cursor specifically may not justify it.

What This Book Assumes

The rest of this book assumes you've decided to use Cursor seriously. If after this chapter you've decided one of the alternatives fits better, that's a fine outcome — pick what fits your work. If you're choosing Cursor or already on it, the rest of the book is about getting the most out of it.

Chapter 3 covers the plans in detail — Hobby vs. Pro vs. Pro+ vs. Ultra vs. Teams vs. Enterprise — so you can choose the one that fits your usage before you start configuring the tool.

Chapter 3: Choosing the Right Plan

Cursor's pricing is the most controversial part of the product. The 2025 transition from request quotas to credit pools generated more user frustration than any other change in the company's history, and the residual confusion about what each plan gets you is real. This chapter walks through the plans honestly, including where each one fits and where each will quietly drain more of your money than expected.

Pricing details and credit allocations change. The numbers here are accurate as of early 2026; check cursor.com/pricing for the current state before you subscribe.

How Cursor Pricing Works

Before walking through the individual plans, understand the underlying model.

Every paid Cursor plan gives you a **monthly credit pool** measured in dollars. When you use the editor, every action that calls a frontier model — Tab autocompletes, Cmd+K edits, chat messages, Manual Edit proposals, Agent runs, Background Agent runs — consumes credits at usage-based prices that vary by model. Tab autocomplete is essentially free because it uses Cursor's own fast model. Claude Opus 4.7 in Agent mode is the most expensive option by far and can drain a Pro plan's credit pool in two or three days of heavy use.

When your monthly credits run out, three things can happen depending on how you've configured your account:

1. **Hard cap (default).** Frontier model access stops until the next billing cycle. Tab and the cheapest models still work; everything else is gated.
2. **Pay-as-you-go.** You opt into usage-based billing beyond your included credits. Your card is charged at the same per-token rates as the included credits.
3. **Budget alert with hard cap.** You set a maximum monthly spend; Cursor warns you when you approach it and stops you at the cap.

The credit pool is the unit of pricing comparison across the plans. Pro includes $20/month of credits. Pro+ includes $60/month. Ultra includes the equivalent of around $400/month worth at usage-based rates (the marketing number is "20x

Pro usage"). Actual mileage depends entirely on which models you use and how disciplined your workflow is.

Tab autocomplete and Cmd+K with the cheapest models are essentially unlimited on every paid plan. The credit pool gates the frontier models — Claude Opus, GPT-5.4, DeepSeek V4 Pro, Gemini 3.1 Pro — and the Agent and Background Agent workflows that use them.

Hobby (Free)

The Hobby tier is what you get when you sign up without paying. It's enough to evaluate the product but not enough to use it seriously day-to-day.

What you get:

- **Tab autocomplete** — limited monthly count, refreshes monthly. Enough for occasional use.
- **Cmd+K and chat** — limited monthly count of "fast" requests using mid-tier models. Frontier models are not available.
- **Codebase indexing** — works on small codebases. Larger projects may hit indexing limits.
- **No Agent mode for premium models.** You can experiment with Agent on the smaller models, but the workflow that makes Agent valuable is gated.
- **No Background Agents, no MCP, no rules-import features beyond the basics.**

Hobby's right fit: students learning to code, developers evaluating Cursor before committing, occasional users who only need AI for simple completions. For real client work or a startup's daily development, you'll outgrow Hobby in days.

Pro ($20/month)

Pro is the plan most working developers should start with. It's where the product becomes useful and where the credit-pool model first matters.

What you get:

- **Unlimited Tab autocomplete** with no monthly cap.

- **Unlimited Cmd+K** with the standard models.
- **$20/month credit pool** for frontier models in chat, Manual Edit, and Agent.
- **Access to all four frontier models** in the picker — Claude Opus 4.7, GPT-5.4, DeepSeek V4 Pro, Gemini 3.1 Pro — when credits allow.
- **Agent mode** with full capabilities, gated by the credit pool.
- **Background Agents** with limited monthly minutes (subject to the credit pool).
- **MCP support** — install MCP servers, use them in Agent.
- **Project Rules** — full `.cursor/rules/` support.
- **Privacy mode** — enable zero-data-retention for sensitive codebases.

Where Pro shines: a single developer working full-time, with discipline about model selection. If you're using DeepSeek V4 Pro or Gemini 3.1 Pro for the bulk of your work and reserving Opus for the gnarly cases, $20/month of credits covers a productive month.

Where Pro hurts: the "Opus on every request, Agent for every task" workflow. A developer who defaults to the most expensive model and runs Agent on tickets that should have been Manual Edit can drain the entire month's credit pool in three to five days. The infamous "$20 in three days" trap is real, and it's the most common pain point for new Cursor users. Chapter 19 walks through the cost discipline that prevents it.

Pro's right fit: solo developers, freelancers, employed engineers paying out of pocket, anyone who wants to try Cursor seriously without committing to a higher tier. Most readers of this book should start here.

Pro+ ($60/month)

Pro+ exists for users who consistently hit the Pro credit cap before the month ends. Same Pro feature set with three times the credit pool — $60/month worth at usage-based rates.

What you get:

- **All Pro features.**
- **3x the monthly credit pool** ($60/month worth at usage-based rates).

- **Higher Background Agent minute allocation.**
- **Modestly faster priority on shared resources** during peak times.

Pro+'s right fit: heavy daily users who consistently hit Pro's cap and don't want pay-as-you-go billing. If you're paying $40/month in usage-based overages on top of your Pro plan, Pro+ is straightforwardly cheaper.

Pro+'s wrong fit: users who haven't built the workflow discipline that makes credit consumption predictable. Upgrading to Pro+ to "buy your way out" of the credit cap before fixing your workflow can mean spending $60/month and still hitting it. The book's advice: try one disciplined month on Pro before assuming Pro+ is necessary.

Ultra ($200/month)

Ultra is the highest individual tier. It exists for power users who treat Cursor as their primary tool and cost is a non-issue.

What you get:

- **All Pro+ features.**
- **20x Pro's credit pool** — roughly $400/month worth of usage at standard rates.
- **Highest priority access** during peak load.
- **Increased Background Agent capacity.**
- **Some early-access features** during staged rollouts.

Ultra's right fit: contractors and freelancers billing AI-assisted work back to clients at rates where $200/month is a small fraction of a single billable day. Engineers running long Background Agent workflows continuously. Solo founders pushing a product hard during a launch sprint. People for whom Cursor's leverage is worth more than $200/month and predictable cost is the value.

Ultra's wrong fit: most working developers. Even heavy users can usually fit comfortably in Pro+ with workflow discipline. If you're considering Ultra "to be safe," try Pro+ for two months first and look at your actual usage. If you're consistently using more than $80/month in credits on Pro+, Ultra is worth a look.

Teams ($40/user/month)

Teams is the first tier that turns Cursor from a personal tool into a team workflow. Pricing is per-user-per-month and includes meaningful organizational features.

What you get per user:

- **All Pro features for each user.**
- **Pooled credit allocation** — credits shared across the team rather than per-user, smoothing out individual spike days.
- **Centralized billing** — one invoice, one payment method, expense reporting for finance.
- **Usage analytics** — admin dashboard showing per-user usage, credit consumption, model preferences.
- **Shared rules** — `.cursor/rules/` libraries managed and distributed centrally.
- **Shared MCP configurations** — admins install and approve MCP servers org-wide.
- **SAML/OIDC SSO** — sign-in via your team's identity provider.
- **Role-based access controls** — admins, members, and limited guest access patterns.
- **Privacy mode enforced org-wide** if your team needs it.

Teams' right fit: any team of 3+ engineers using Cursor seriously. Shared rules alone justify the upgrade — pushing rule changes through a central location instead of asking everyone to update their local config saves hours per month per team. Pooled credits also smooth out the "one engineer's expensive day spikes the bill" problem on individual plans.

Teams' wrong fit: a single engineer or two who don't need shared configuration. Stay on Pro+ until team size justifies it.

The book's recommendation: small teams should plan to move to Teams within their first month of serious adoption, not wait until they hit a specific feature need. The shared rules workflow compounds over time and starts paying off immediately.

Enterprise

Enterprise is the negotiated-pricing tier for large organizations. It's not self-serve; you talk to Cursor's sales team and they build a contract for your needs.

What you typically get:

- **All Teams features for each user.**
- **SCIM provisioning** — automated lifecycle management with your IdP.
- **Audit logs** — comprehensive activity logging for compliance.
- **Custom contractual terms** — DPA, MSA, SLA, custom data-residency clauses.
- **SOC 2 Type 2 attestation** — for your security team's review.
- **Repo, model, and MCP blocklists** — administrators can restrict which repositories Cursor can index, which models users can call, and which MCP servers can be installed.
- **AI-code-tracking API** — programmatic visibility into AI-generated code in your repos.
- **Pooled usage billing** — invoice billing rather than card billing.
- **Dedicated support** with named CSM and engineering escalation paths.
- **Privacy mode and zero-data-retention** as default policies, enforceable org-wide.

Enterprise's right fit: any organization with formal security review processes, regulated workloads, or a need for SCIM and audit logs. If your security team needs to threat-model the tool before approving it, you're in Enterprise territory.

Enterprise's wrong fit: small teams that don't need the compliance features. Teams already provides SSO and centralized billing, which covers many small companies.

Pricing is negotiated and ranges substantially. Expect a multi-month sales cycle, a security review, and a per-user price that can be lower or higher than the Teams rate depending on volume and term. Chapter 28 covers the enterprise rollout in more detail.

BugBot — A Separate Add-On

BugBot is Cursor's automated PR reviewer-and-fixer (covered in detail in Chapter 23). It's priced separately from the editor subscription:

- **BugBot Pro** — usage-based pricing, integrated with your existing Pro/Pro+/Ultra subscription.
- **BugBot Teams** — included with Teams subscriptions in most cases, with usage limits.
- **BugBot Enterprise** — included with Enterprise contracts, with custom SLAs.

BugBot's value depends on your team's PR volume and review culture. A team that ships 50+ PRs per week and reviews every one will get substantial value. A two-person team merging a few PRs a week may not need it. Chapter 23 covers the cost-benefit analysis.

A Decision Flowchart

A rough flowchart for choosing among the plans:

Evaluating Cursor for the first time? → Hobby for a few days. Don't pay before you know you'll use it.

Working developer planning to use Cursor seriously? → Pro at $20/month. Stay here for at least one full month before considering upgrades.

Pro plan running out of credits before the month ends, and you've already tightened model selection? → Pro+ at $60/month, or pay-as-you-go on top of Pro if you want the metered visibility.

Usage consistently above $80/month in credits and you want predictable budgeting? → Ultra at $200/month.

Team of 3+ engineers using Cursor seriously? → Teams at $40/user/month.

Organization requires SCIM, audit logs, or custom security/compliance terms? → Enterprise. Talk to sales.

Reviewing 30+ PRs per week and want automated review? → Add BugBot to whichever editor plan you're on.

Common Mistakes in Plan Selection

A few patterns the book flags explicitly, because they cost real money:

Buying Ultra "to be safe" without measuring actual usage. Ultra is expensive. Most users who think they need it actually need Pro+ with better model discipline. Try Pro+ first.

Staying on Pro long after the team has grown. Five engineers each on individual Pro plans pay $100/month in subscriptions and lose the shared rules benefit. Teams at $200/month costs more in absolute terms but pays for itself in workflow leverage and centralized administration.

Using Hobby for "real" work. Frequent friction from hitting Hobby's limits costs more than the $20/month for Pro, even if it doesn't show up on a credit card. Time fighting limits is time not shipping.

Buying Enterprise before the company is ready. Enterprise requires a security review, contract negotiation, and operational maturity. A 10-person startup probably doesn't need it yet — Teams is the right tier until you have formal compliance requirements.

Ignoring BugBot pricing. BugBot's pricing adds to the editor plan. Add it to your budget calculations explicitly; teams sometimes get surprised by the combined invoice.

From Plan to Editor

You've picked a tier. Chapter 4 walks through installation on Mac, Windows, and Linux, the settings panel, and the first-launch decisions that matter — including Privacy mode, the most consequential configuration choice for users working on sensitive codebases.

Chapter 4: Installing and Configuring Cursor

This chapter walks through getting Cursor onto your machine, importing your existing VS Code setup, signing in, and making the configuration choices that matter at first launch. Most of it is straightforward. The parts that aren't — Privacy mode, model selection, indexing on large codebases — get extra attention here because the defaults aren't always right for your context.

Downloading and Installing

Cursor is distributed at cursor.com. The download page detects your operating system and offers the right installer. The download is roughly 200 MB.

macOS. A signed and notarized `.dmg` for Apple Silicon and Intel. Drag Cursor into your Applications folder, launch it, and on first run macOS will ask whether you trust the application. The installer is universal — the same `.dmg` works on M-series and Intel Macs.

Windows. A signed `.exe` installer. Run it, click through the wizard, choose whether to add Cursor to your PATH (yes — you'll want the `cursor` CLI command), and let it complete. A portable ZIP is also available if you can't run the installer.

Linux. An AppImage and a `.deb` package, plus a tarball for distributions that prefer manual installation. The AppImage is the lowest-friction option for Ubuntu, Fedora, and most other mainstream distributions. Make it executable with `chmod +x` and run it. Community-maintained packaging exists in some distribution package managers, but the official AppImage and `.deb` are the supported paths.

After installation, Cursor prompts you on first run to either sign in or create a new account.

First Launch and Account Setup

The first launch sequence asks you to make several decisions. Most have sensible defaults, but two — VS Code import and Privacy mode — are worth paying attention to.

Sign in or create an account

Sign in with Google, GitHub, or email-and-password. If your team has SSO configured (Teams or Enterprise), you'll typically sign in through your team's identity provider. Sign-in opens a browser window for OAuth authentication and returns you to the editor on success.

The book's recommendation: use the same identity you use for GitHub, since Cursor's GitHub integration is one of the more useful features and starting with a different identity creates friction later.

Import VS Code settings

If you have VS Code installed, Cursor offers a one-click import of your settings, extensions, and keybindings. Take the import. It's almost always the right choice — themes, tab settings, fonts, favorite extensions, custom keybindings all transfer in seconds.

Two caveats:

- **Some extensions don't transfer cleanly.** Mostly the ones depending on VS Code-specific Microsoft proprietary marketplace components. The vast majority install without issue, but you may need to manually replace one or two from the Open VSX registry that Cursor uses.
- **You can re-run the import later.** If you skip it now or want to refresh after a VS Code change, the command palette has "Cursor: Import VS Code Settings" available at any time.

Choose your editor theme and font

Identical to VS Code: a theme picker (or the imported VS Code theme), font family, font size. If you imported VS Code settings, this is already done.

Decide on Privacy Mode

This is the most consequential choice on first launch and the one users most often regret getting wrong.

Privacy mode off (default). Cursor's servers receive your code as part of normal model invocations. Cursor uses the data per its standard data handling — covered at cursor.com/privacy — which works for most use cases but isn't zero-data-retention.

Privacy mode on. Zero-data-retention infrastructure. Code touches Cursor's servers in transit but isn't stored beyond the request. This is the right setting for any work on proprietary codebases, regulated workloads, or anything where your organization's policy requires it.

The book's recommendation: turn Privacy mode on at first launch unless you have a specific reason not to. The latency cost is negligible and the privacy posture is meaningfully better. Chapter 21 covers Privacy mode's threat model in detail; if you're working in a regulated environment, read that chapter before deciding.

For Enterprise users, Privacy mode is typically enforced at the org level and may not be a per-user choice.

The Settings Panel

Cursor's settings live in the same Command Palette → "Preferences: Open Settings" path as VS Code, plus a Cursor-specific settings panel accessed via the gear icon or Cmd+, (Ctrl+, on Windows/Linux). The Cursor-specific settings cover the AI primitives VS Code lacks.

The most important sections at first launch:

Account and billing

Shows your current plan, credit balance, usage history, and billing settings. This is where you set up pay-as-you-go usage if you want to opt in, configure a hard spending cap, or upgrade your plan.

The book's recommendation: set a hard cap your first month even if you don't think you'll hit it. The "$20 in three days" trap is real for new users, and waking up to a $300 charge because you forgot you opted into usage-based billing is a bad day.

Model selection

The default model picker determines which model Tab autocomplete, Cmd+K, chat, Manual Edit, and Agent use unless you override per-request. The picker offers Auto (Cursor selects per task), the four frontier models (Claude Opus 4.7, GPT-5.4, DeepSeek V4 Pro, Gemini 3.1 Pro), and the cheaper model tiers.

The book's recommendation: leave the default at Auto for the first week. Auto picks reasonably well across tasks, and overriding to a specific model before you've built intuition for which model fits which task tends to either underspend (always picking the cheap model) or overspend (always picking Opus). Chapter 18 covers the picker in detail.

Codebase indexing

The settings panel has an indexing section showing which projects are indexed, the indexing status, and per-project ignore configuration. We cover indexing in Chapter 13. At first launch, leave the defaults — when you open a project, Cursor will offer to index it.

Privacy mode

If you didn't enable Privacy mode at first launch, you can change it in settings at any time. The change takes effect immediately for new requests.

Tab autocomplete behavior

Tab settings include trigger sensitivity, when to suggest, multi-line aggressiveness, and language-specific overrides. The defaults are good. Don't tune Tab until you have at least a few weeks of usage to inform what you'd want different.

Keybindings

If you imported your VS Code keybindings, this section shows them. Cursor adds a few keybindings on top — Cmd+K for inline edit, Cmd+L for chat, Cmd+I for Manual Edit, Cmd+Shift+I for Agent. The defaults are sensible; if you have conflicts with your existing keybindings, resolve them here.

Installing Extensions

Cursor uses the Open VSX registry for extensions, not the Microsoft VS Code Marketplace. This is a Microsoft licensing requirement, not a Cursor design decision. The practical difference: most popular extensions are available on Open VSX, but some publisher-specific extensions (typically Microsoft's own, like Pylance) may not be. Workarounds for getting Microsoft-only extensions in Cursor are documented at cursor.com/docs, but they're outside the book's scope and most users won't need them.

Open the Extensions panel (Cmd+Shift+X / Ctrl+Shift+X), search for the extensions you want, click Install. Imported VS Code extensions install automatically during the first-launch import, where available.

Setting Up Git Integration

Cursor uses VS Code's Git integration unchanged. If you've used Git in VS Code, it works the same way in Cursor. The status bar shows your branch and changes; the Source Control panel (Cmd+Shift+G / Ctrl+Shift+G) handles staging, committing, and pushing.

Cursor adds AI-assisted commit messages: in the Source Control panel, a sparkle icon proposes a commit message based on your staged changes. The proposed messages are usable but bland; most teams prefer their own conventions and keep this feature disabled. Turn it off in settings under "Cursor → Git → AI Commit Messages."

Connecting GitHub for Background Agents and BugBot

Background Agents and BugBot both depend on a GitHub connection for repository access. Cursor prompts you to authorize the GitHub App when you first try to use those features, but you can do it preemptively in Settings → Integrations → GitHub.

The GitHub App requests permissions to:

- Read repository content (required)
- Create and update branches (for Background Agent commits)
- Create pull requests (for Background Agent and BugBot output)

- Read and write PR comments (for BugBot review and replies)

If you're a member of multiple GitHub organizations, the authorization flow lets you scope access per-organization. The book's recommendation: scope tightly. Authorize Cursor only on the organizations where you'll use Background Agent and BugBot. Extend later if needed.

For Enterprise users, the GitHub integration is often configured at the organization level rather than per-user, with an admin granting org-wide access through GitHub Enterprise's standard app installation flow.

Configuring `.cursor/` for the First Project

Most of Cursor's per-project configuration lives in a `.cursor/` directory at your project root. The directory is meant to be checked into version control — it's project configuration, not user state — and shared across the team.

The minimum starter `.cursor/` looks like this:

```
.cursor/
├── rules/
│   └── general.mdc        # Project-wide rules
├── ignore                 # Files Cursor should not index
└── environment.json       # (Optional) Agent execution environment
```

Don't worry about creating this directory at first launch — Cursor creates it the first time you write a rule or configure an ignore list. For now, it's enough to know that the directory exists and that the configuration in it is per-project, not per-user. Chapters 14, 15, and 25 cover the contents in detail.

Verifying the Install

A quick sanity check after install:

1. Open the command palette (Cmd+Shift+P / Ctrl+Shift+P) and run "Cursor: Show Logs." If logs are present and recent, the editor is talking to Cursor's services.
2. Open a small project and try Cmd+K on a function: select the function, hit Cmd+K, type "add a docstring," and confirm a proposed diff appears. If yes, models are working.

3. Open the chat panel (Cmd+L) and ask a simple question about your project. Confirm a response comes back.
4. Type some code and watch Tab autocomplete propose a completion. Confirm the proposal is reasonable.

If all four work, you're set up. If any fail, the most common causes are network issues (corporate proxies blocking Cursor's API endpoints), authentication issues (an expired session), or extension conflicts (a VS Code extension misbehaving). The Cursor docs have a troubleshooting page at cursor.com/docs/troubleshooting that covers the common cases; Appendix C covers the most frequent issues we've seen in practice.

A Note on Updating Cursor

Cursor auto-updates by default. Updates ship frequently — sometimes multiple times per week — and they're usually small bug fixes and incremental improvements, occasionally with new features. Auto-update is fine for most users.

If you're on Enterprise or Teams and need version control over the update cadence (some compliance regimes require it), Cursor supports pinning to specific versions and rolling out updates manually. This is configurable in Enterprise admin settings.

For individual users: leave auto-update on. The friction of managing updates manually isn't worth the marginal control.

With a working editor, an account, and your VS Code settings carried over, you're configured. Chapter 5 picks up with the first edits — Tab autocomplete, Cmd+K, the chat sidebar, and ten useful starter prompts that exercise different parts of the daily rhythm before Part II dives into the three modes.

Chapter 5: Your First Edits with AI

This chapter walks you through your first hour of real use. Tab autocomplete, Cmd+K inline edits, the chat sidebar, ten starter prompts that exercise different parts of the editor, and a small refactoring walkthrough on a real file. By the end you should have a feel for how Cursor's daily rhythm differs from your previous editor.

The chapter is short on theory and long on doing things. Theory comes in Parts II and III; right now we want you typing and watching the editor respond.

Tab Autocomplete

Tab is the simplest AI feature in Cursor and the one you'll use most often. As you type, Cursor's Tab model proposes completions inline in a faded color. Press Tab to accept; keep typing to ignore.

Three things to notice in your first hour with Tab:

It uses your codebase, not just the current file. When you start typing a function call, Tab can complete it with arguments that match how the function is used elsewhere in your project. When you start writing a test, Tab can complete it in the style of nearby tests. The codebase awareness shows up most clearly when you compare it to AI features that only see the current file.

It can rewrite as well as complete. Cursor's Tab model can propose edits to lines you've already written when it detects an obvious change you'd want — adding a missing import after you reference it, fixing a typo in a variable name you've used consistently, completing the matching brace in a long structural change. These rewrites appear as multi-line proposals; you accept the same way (Tab) and they apply across the proposed range.

It's wrong often enough that you have to read it. Tab's accuracy is high — somewhere in the 50–70% useful range for most users — but not perfect. Train yourself early to read the proposed completion before accepting. The habit of accepting Tab without reading starts the bad-code spiral Chapter 30 covers in more detail.

Try it now: open a file in any project, start writing a function, watch Tab propose. Accept some, reject some, get a feel for when it's useful and when it isn't.

Cmd+K — Inline Generation

Cmd+K (Ctrl+K on Windows/Linux) is the second feature you'll use constantly. It opens an inline prompt at your cursor or selection.

The basic flow:

1. Place your cursor where you want a change, or select an existing block of code.
2. Press Cmd+K.
3. Type a prompt in the inline box that appears.
4. Press Enter and watch the proposed change appear inline as a diff.
5. Accept the diff (Cmd+Enter) or reject it (Esc).

Some patterns that work well with Cmd+K:

- **"Generate a function that..."** — with no selection, Cmd+K writes new code at the cursor position.
- **"Refactor this to..."** — with a selection, Cmd+K rewrites the selection per your instruction.
- **"Add a docstring"** — with a function selected, generates a docstring that matches the function's behavior.
- **"Convert this to async"** — with a function selected, rewrites it as an async version.
- **"Write tests for this"** — with a function selected, generates a test block (sometimes inline, sometimes prompts to create a test file).

Cmd+K is best at scoped, well-defined edits. If you find yourself writing a long Cmd+K prompt that describes multi-file changes, you've reached the point where Manual Edit or Agent is the better tool. Chapter 11 covers Cmd+K patterns in more depth.

The Chat Sidebar

Cmd+L (Ctrl+L) opens the chat sidebar. This is the most familiar AI surface for anyone who's used ChatGPT or Claude — a conversation pane where you ask questions and get text answers.

In Cursor, the chat sidebar is more powerful than a generic chat:

- **It sees your project.** Without you attaching files, the chat has access to your codebase index and can answer questions about how parts of your project work.
- **You can attach files and folders.** The @ symbol in the chat input lets you reference specific files, folders, symbols, web pages, and documentation. Attachments scope the chat's context to those references.
- **It can propose edits.** Ask the chat to change something and it can propose a diff you accept directly from the chat panel.
- **It runs in any of three modes.** The mode selector at the bottom of the chat panel switches between Ask (read-only), Manual (proposes edits), and Agent (executes autonomously). We cover all three in Part II.

For your first hour, leave the mode on Ask and have conversations about your code. "Explain how authentication works in this project." "Where is the rate limiter defined?" "What are the public APIs of this module?" Get a feel for how it answers.

Ten Useful Starter Prompts

Ten prompts that exercise different parts of Cursor and reveal different capabilities. Try them on a real project.

1. "Walk me through how authentication works in this project." A project-wide question that exercises the codebase index. The answer should reference real files and real code, not generic auth patterns.

2. "What does this function do?" (with a function selected, Cmd+L) — Tests local context awareness. The explanation should match the actual code, not a generic description.

3. "Add a docstring to this function." (Cmd+K with a function selected) — Tests inline generation on a small, scoped task. The docstring should reflect actual parameters and return values.

4. "Write a unit test for this function." (Cmd+K) — Tests test-generation. Watch where the test ends up — inline, in a new file, or in an existing test file matching project conventions.

5. "Convert this from a class component to a function component." (Cmd+K, with a React component selected) — Tests refactoring on a structural change. Check whether hooks usage is correct.

6. "Find any places this configuration value is hardcoded that should be using the environment variable instead." (Cmd+L Ask mode) — Tests cross-file search and analysis.

7. "Why might this test be flaky?" (Cmd+L with a flaky test attached) — Tests reasoning over a specific question. The answer should reference actual test code, not generic flakiness causes.

8. "Refactor this 200-line function into smaller functions, preserving behavior." (Cmd+K) — Tests larger refactor on a single file. Check whether the refactor preserves edge cases.

9. "What's the difference between these two implementations?" (Cmd+L with two files attached) — Tests comparison reasoning across files.

10. "Explain this codebase to a new engineer joining the team." (Cmd+L Ask mode) — Tests big-picture synthesis. The answer should be specific to the project, not generic.

If most of these produce useful answers, Cursor is working well in your project. If most produce vague or off-base answers, the project's index probably hasn't completed — give it ten or twenty minutes and try again. Indexing on a fresh project takes time, especially in monorepos. Chapter 13 covers indexing in detail.

A Walkthrough: Refactoring a File

A small concrete example to ground all of this. The goal: take a real file in a real project and clean it up using Cursor's primitives.

Pick a file in a project of yours that's a little messy — long functions, duplicated logic, missing docs. A 100–300 line file is ideal for this walkthrough; bigger files we come back to in Part IV.

Step 1: Understand the file

Open the file. With nothing selected, open chat (Cmd+L) and type:

> "Walk me through what this file does, function by function."

Read the answer carefully. If it matches your understanding, the model has the file's context loaded correctly. If it's off, double-check that you've selected the right file or that indexing is complete.

Step 2: Find the issues

Same chat session, follow up:

> "What are the three biggest issues you see in this file? Be specific — name the function and the line range."

The model should give concrete pointers. "The `processOrder` function on lines 45–110 mixes validation logic, business logic, and database calls without clear separation." That kind of answer is useful. "The code could be more modular" is not — push back if you get vague answers.

Step 3: Plan the refactor

Same session, follow up:

> "If we were to fix the first issue, what would the refactor look like? Don't write code yet — describe the structural change."

A short plan, written out before the diff, is the difference between a refactor that lands cleanly and one you have to revert. Read the plan, push back if it's wrong, iterate until you agree.

Step 4: Execute the refactor

Switch chat mode to Manual Edit. Type:

> "Implement the refactor we just discussed."

Manual Edit proposes a multi-file diff. Read every line. Accept what's right, reject what's wrong, ask follow-ups for what's confusing. Don't accept a Manual Edit diff because the model "seems confident" — read it.

Step 5: Verify

Run your tests. Hopefully they pass. If they don't, the diff broke something — go back to Manual Edit, describe the failure, and let it fix forward.

This five-step rhythm — understand, find issues, plan, execute, verify — is the core daily rhythm of effective Cursor use. Every part of the rest of the book elaborates on it. Right now, the goal is to feel the cycle in a small task before scaling it up.

When to Use What

A rough mapping for your first week:

- **Single-line changes while you're typing:** let Tab handle them.
- **Selected-block changes, well-defined:** Cmd+K.
- **Questions about the codebase:** chat in Ask mode.
- **Multi-file edits where you've planned the change:** chat in Manual Edit mode (covered in Chapter 7).
- **End-to-end tickets where you want the editor to do the work:** chat in Agent mode (covered in Chapter 8).

This mapping gets refined in Part II. The heuristic for now: smaller and more scoped → Tab or Cmd+K. Bigger and less defined → chat. The chat mode (Ask, Manual, Agent) escalates with the scope of the change.

Common First-Day Mistakes

Three patterns I've seen new Cursor users fall into on their first day:

Accepting Tab without reading it. Easy to do because Tab is fast. Hard habit to break later. Read every Tab proposal before accepting, even when it's "just a one-liner." The wrong one-liner is still wrong.

Using Agent for everything. Agent is impressive on demos. In daily use, most of your wins come from Tab and Cmd+K, with Manual Edit for bigger structural changes and Agent reserved for the autonomous tasks. Defaulting to Agent is the fastest way to drain credits.

Writing huge prompts. A long, detailed prompt for what should have been a five-word ask is a tell that you don't trust the model – and reading the diff carefully beats over-specifying upfront. Keep prompts short for small tasks; reserve detail for the cases where it matters.

Part II: The Three Modes

"Choose the smallest tool that will finish the work. The big one breaks more than it fixes."

— A working principle

Chapter 6: Ask Mode — Read-Only Exploration

Ask is the smallest of the three chat modes and the one most experienced Cursor users underuse. It does one thing: answer questions about your codebase without changing anything. No file edits, no terminal commands, no proposed diffs. Just reading and reasoning.

Sounds limited. In practice, it's the mode that protects you from the most common failure in AI-assisted development — making changes before you understand the code. This chapter is about why Ask matters, when to reach for it, and the workflow patterns that turn it into a daily habit rather than a forgotten feature.

What Ask Mode Does

When you open the chat sidebar (Cmd + L / Ctrl + L) and select Ask in the mode dropdown at the bottom of the panel, you're in a read-only conversation. The model has access to your codebase index, can read any file you reference, can search across files, and can fetch web pages and documentation if you give it a URL or invoke a search. It cannot:

- Edit files in your project
- Run terminal commands
- Modify the filesystem
- Open pull requests or push commits

Everything else a chat conversation does, Ask does — showing you code blocks you can copy, walking through reasoning, comparing implementations, suggesting changes you might want to make. It just can't make those changes itself.

This is the point. Ask is for the part of the work that comes before you decide what to change.

What Ask Mode Doesn't Do

A few common confusions worth clearing up:

Ask still proposes code; it just doesn't apply it. Ask "how should I rewrite this function" and Ask writes you the rewritten function in a code block. Copy it manually or switch to Manual Edit/Agent mode and ask the same thing to apply it. The code Ask suggests is exactly what Manual would propose — the difference is the apply step.

Ask is not slower than the other modes. The misconception is that Ask is somehow lighter-weight because it doesn't change files. The model's reasoning work is the same; only the output channel differs. A complex Ask question on Opus 4.7 costs the same in credits as the equivalent Manual Edit query on the same model.

Ask can't refuse to make changes you ask for. It can't make them in the first place. There's no "Ask is being cautious"; it's structurally read-only. Ask "can you actually edit this," and the model will tell you to switch modes.

Ask doesn't remember between sessions any more than the other modes. Cursor preserves your chat history per-project, but Ask isn't a special long-term memory store. Project Rules and the codebase index persist; chat history is conversational state.

When Ask Is the Right Tool

Five situations where Ask is the right choice — and where reaching for Manual or Agent is actively the wrong move:

You don't yet understand the code. The single most common cause of bad AI-assisted edits is the user not having a clear picture of what the existing code does before changing it. Ask is the answer. Spend ten minutes asking questions until you have a working mental model, then switch modes. The ten minutes saves you from forty minutes of debugging the model's well-intentioned wrong fix.

You're scoping a change. Before you write a ticket, before you plan a refactor, before you decide whether a piece of work is small or large — Ask. "How big is the change to add field X to the user record? What touches the user record today? Are there any indexed paths I should be aware of?" The conversation produces a scope estimate Manual or Agent can act on. Skipping the scoping step and jumping straight to "implement field X" is the path to PRs that touch twice as many files as needed.

You're drafting a design doc or PR description. Ask shines for synthesis tasks — write a one-pager explaining how authentication works, draft release notes for these last twenty commits, write a PR description for these changes. The output is text you copy out, not code that needs to be applied. No diff to review; the friction of Manual mode for a pure-text task is overhead.

You're investigating a bug without enough information yet. The instinct in a debugging session is to start trying fixes. The better workflow is to start by asking — "where could this null come from? what writes to this field?" — until you have a hypothesis, then switch to Manual or Agent to test the fix.

You're explaining the codebase to someone else. Onboarding a new engineer? Writing internal docs? Explaining the architecture to a stakeholder who doesn't read code? Ask the chat to walk through the relevant parts and summarize. The output is meant for humans to read, not for the editor to apply.

When Ask Is the Wrong Tool

Two situations where Ask is just stalling:

You already know what you want to change and you're "just gathering context." If the change is well-scoped in your head, jump to Manual Edit. Asking the model to confirm what you already know is busywork that doesn't change the outcome.

You're asking Ask to make changes and copy-pasting them by hand. If the conversation is producing diffs you apply manually, switch to Manual mode and let the editor apply them. Manual gives you the same diff in a reviewable format with one-click apply. Copy-paste is slower and more error-prone.

The line between "good Ask use" and "stalling" is whether the conversation changes your mental model or whether you're using Ask as a less-functional Manual.

Walkthrough: Understanding an Unfamiliar Codebase

Ground this in a concrete scenario. You've just been added to a project — say, a payments service you didn't write. The team wants you to add support for a new payment provider next week. Where do you start?

Open the project. Let it index for a few minutes. Open chat (Cmd+L), set mode to Ask, and walk through these prompts in order. Adapt them to your project specifics.

1. The architectural overview.

> "Walk me through the high-level architecture of this codebase. What are the main layers, what's the request flow for a payment, and where does external-provider integration live?"

Read carefully. If it matches what someone described in onboarding, the model has a real picture of the project. If it's vague or generic, the index isn't ready or something unusual is going on with the project's structure. Don't proceed until the answer is concrete.

2. The existing provider integration.

> "Show me where the existing Stripe provider is integrated. Trace the call path from incoming request to outbound provider call. What interface or abstraction does Stripe implement?"

This identifies the seam where the new provider plugs in. The model should reference real files and real interfaces. Push back if it's vague.

3. The shared abstractions.

> "Is there a shared interface, base class, or set of utility functions that all providers use? Where are tests for the existing provider, and what patterns do they follow?"

This answers "what's the implementation cost." Clean abstraction layer? Cost is low. Stripe wired in directly with no abstraction? Cost includes building the abstraction first.

4. The edge cases the existing code handles.

> "What edge cases does the Stripe integration handle today? Webhook idempotency, retries, partial failures, currency mismatches — what does the existing code do for each?"

This is the list of things your new provider must also handle. Skip this and you'll ship a provider that works on the happy path and fails on the edge cases the existing code learned about over years.

5. The known weaknesses.

> "Looking at the Stripe integration, what's the weakest part of the design? If you were going to refactor anything, what would it be?"

The model's answer is your warning system. If the existing integration has a quirk, you'll inherit it when you copy the pattern. Knowing the quirk in advance lets you decide whether to copy it, fix it, or design around it.

6. The plan.

> "Given everything we've discussed, what's a reasonable plan to add a new provider? Outline the work as a sequence of small PRs, each individually shippable."

Ask synthesizes the conversation into a plan. Read it. Push back on anything wrong. The plan you agree on at this point is the ticket you'd write for a junior engineer — and it's the input to Manual Edit or Agent when you start implementing.

This walkthrough takes 15–30 minutes and replaces hours of slow ramp-up. The credits cost is real but small compared to the time saved. The output — a clear plan validated against the actual code — is what makes the implementation phase fast.

Drafting Plans and Design Docs

Beyond exploration, Ask is the right tool for planning artifacts that are text, not code. Some patterns:

The implementation plan. "Outline the work to migrate from REST to gRPC for the user service. Identify the steps, the rollback plan, and the risks. Format as a numbered list."

The design doc. "Write a design doc for adding webhook support to this API. Include goals, non-goals, proposed design, alternatives considered, and a rollout plan. Reference the existing patterns in the codebase."

The PR description. "Draft a PR description for the changes I'm about to make to the rate limiter. Include the motivation, the approach, the testing strategy, and a checklist for reviewers."

The handoff doc. "Write a doc for the engineer who will own this code after I leave the team. Cover the architecture, the gotchas, where the surprising decisions were made and why, and what to be careful about during changes."

For all of these, Ask's output is meant to be edited by you, not by the editor. Read carefully, rewrite where the tone or specifics are wrong, copy into your destination. The model is a strong first-draft writer for technical text; it's not a stronger writer than you are once you've done a careful pass.

Reading the Codebase With @ References

The chat input supports @ references that scope what the conversation can see. The references work in all three modes; in Ask mode they're useful for narrowing the model's attention to the right files when the codebase is large.

Common @ patterns:

- `@filename.ts` — attaches a single file. The model reads the full file as context.
- `@folder/` — attaches a folder. The model can see all files in the folder; in large folders this gets summarized rather than fully loaded.
- `@SymbolName` — attaches a function, class, or other named symbol. The model gets the symbol's definition and immediate context.
- `@Codebase` — explicitly invokes a codebase-wide retrieval. Useful when your question crosses many files.
- `@Web` — invokes web search. Useful for questions about external libraries, APIs, or current best practices.
- `@Docs` — references the documentation sources you've configured for the project.

Layering @ references is how you get focused answers. "Looking at @PaymentProcessor and @StripeAdapter, what's the cleanest way to add a new provider?" gives the model exactly the context it needs without burning tokens on unrelated files.

The book's recommendation: when an Ask answer is vague, the fix is more @ references, not a longer prompt. A focused prompt over the right files beats a long prompt over the wrong context.

A Mode-Switching Habit

Ask works best as part of a mode-switching habit, not as a destination. The pattern:

1. **Start in Ask.** Build the mental model. Validate the plan.
2. **Switch to Manual.** Implement the changes you've planned.
3. **Switch back to Ask** when you hit something you don't understand mid-implementation.
4. **Repeat.**

The chat panel preserves history across mode switches, so the conversation stays coherent. Ask in Ask, switch to Manual, ask the same chat to implement, then switch back to Ask to investigate a follow-up question — all in one continuous session.

This rhythm is the daily texture of effective Cursor use. Engineers who never switch out of Agent burn credits and ship messy code. Engineers who never use Ask waste time changing code they don't understand. Engineers who fluently move between the three modes get the leverage the tool offers.

You've seen Ask mode and know when reaching for it produces real value versus when it becomes a way to avoid committing to a change. Chapter 7 covers Manual Edit — the multi-file proposal-and-review workflow that's the most precise of the three modes when you already know what you want. Chapter 8 covers Agent, where the editor takes the wheel. Chapter 9 closes Part II with the decision rubric for choosing among them.

Chapter 7: Manual Edit Mode (formerly Composer)

Manual Edit is Cursor's middle gear. The model proposes coordinated changes across one or many files, you review the proposed diffs, and you accept or reject — file by file, hunk by hunk, line by line if you want. The model doesn't run terminal commands or modify files until you say so. The diff is the entire interface.

This is the mode that does the work for the bulk of well-scoped engineering tasks. Agent gets the demos and the headlines; Manual gets the actual day-to-day. The Cursor engineers I've talked to estimate 60–70% of their own daily Cursor use is in Manual mode, with Tab and Cmd+K filling the small-edit space and Agent reserved for the autonomous tasks.

This chapter is about getting fluent in Manual: when to reach for it, how to write prompts that produce reviewable diffs, how to read the diffs critically, and the patterns that separate "Manual edits that ship cleanly" from "Manual edits that introduce subtle bugs."

What Manual Edit Does

When you open chat (Cmd+L) or use the dedicated Manual Edit shortcut (Cmd+I / Ctrl+I) and select Manual mode, you're in a workflow where:

- You describe the change you want
- The model proposes a coordinated set of edits across one or more files
- The proposed edits appear as a diff tree in a review panel
- You review the diffs — accept all, reject all, or pick through file-by-file
- Accepted edits apply to your filesystem; rejected edits leave the file unchanged

The model can read your codebase index, search across files, and reference @-attached files for context. What it cannot do in Manual mode:

- Run terminal commands (no `npm install`, no `pytest`, no shell)

- Modify files outside the proposed diff (you see everything that changes)
- Take multiple turns autonomously — each round is one proposal you review

That last point matters. Agent mode loops — proposes, executes, reads results, proposes more. Manual is single-turn from the model's perspective: it proposes one diff, you act, then if you want more it's a new round.

The Composer Naming

Some readers will recognize the workflow as Composer, the original name for the feature when it shipped in late 2024. Cursor renamed Composer to Manual Edit in 2025 as part of consolidating the three-mode model. The capability is the same; only the label changed. Most internal Cursor docs and the community still use "Composer" interchangeably, and you'll see both in the wild.

The book uses "Manual Edit" or just "Manual" throughout, but if you read older blog posts or YouTube tutorials about "Composer," that's this mode.

When Manual Is the Right Tool

Manual is the right answer when three conditions hold:

You know what you want to change. The change is well-scoped in your head. You can describe it in a sentence or two without hand-waving. "Rename `getCwd` to `getCurrentWorkingDirectory` everywhere." "Add an optional `timeout` parameter to all the public functions in this module, defaulting to 30 seconds." "Convert these three React class components to function components."

The change touches more than a few lines. Single-line and small-block changes belong to Tab and Cmd+K. Manual is for when you'd otherwise be opening four files and editing the same pattern in each. The cost-benefit of switching from Tab/Cmd+K to Manual breaks even around three coordinated edits in different places.

You don't need the editor to verify the change for you. If implementing the change requires running tests, executing scripts, or doing something the editor needs to react to, you want Agent mode. Manual proposes; it doesn't execute. If you'll need to run something to know whether the change worked, Agent's loop is the right tool.

A few concrete cases where Manual is the obvious choice:

- **Multi-file renames** the IDE's built-in rename refactor doesn't handle (because the symbol is referenced via string, configuration, or some other indirection the symbol-based refactor misses).
- **Adding an optional parameter throughout a module** when the parameter has a non-trivial default that varies by call site.
- **Converting a feature to use a new dependency** — replacing every call to library A with the equivalent in library B, when the mapping is consistent but mechanical.
- **Coordinated changes across the boundary between two files** — adding a new field to a model, exposing it through the API layer, plumbing it into the frontend types, all in one diff for atomic review.
- **Style/convention enforcement** — applying a Project Rule across many files when a linter can't express the rule.

For all of these, Manual's win over Tab+Cmd+K is the atomicity of review. You see the whole diff at once and decide whether the coordination is correct, instead of accepting twenty small Tabs and hoping the cumulative result is consistent.

When Manual Is the Wrong Tool

Manual is the wrong tool in three cases:

You don't know what you want to change. You're still scoping. Switch to Ask. A vague Manual prompt produces a vague Manual diff that covers more or less ground than the actual task and either over-scopes or under-scopes the change.

The change is small enough for Cmd+K. A single-block edit on one file is what Cmd+K is for. Reaching for Manual on a one-file change adds the overhead of the diff-review panel for no gain.

The change requires running things. "Add a new dependency and install it." "Run the test suite and fix failures." "Execute the migration and verify it." These are Agent's territory. Manual proposes the change but won't run the install, the tests, or the migration. You'd end up implementing in Manual, then manually doing the execution work — at which point you should have been in Agent.

Writing Good Manual Prompts

The prompt is the input to a single round of proposal-and-review. The diff is the output. The quality of the diff is downstream of the quality of the prompt.

A good Manual prompt has three components:

1. The scope. What files or what code. "In `auth/middleware.ts` and `auth/session.ts`..." or "Across all files that import from `legacy-utils`..." gives the model the boundary. Without it, you're hoping the model picks the right scope from your verbal description.

2. The change. Concrete and specific. "Rename `verifyToken` to `validateAuthToken`" not "improve the auth code." "Add a `timeout: number = 30` parameter" not "make this configurable." The model executes specific instructions well and infers vague intent badly.

3. The constraints. What not to change, what behavior to preserve, what tests to keep passing. "Preserve all error messages exactly. The test in `__tests__/auth.test.ts` should continue to pass without modification." Constraints reduce surprise in the diff.

A complete example:

> "In `src/payments/stripe.ts` and `src/payments/types.ts`, rename `processCharge` to `submitCharge` everywhere it appears, including in JSDoc comments. Update the type exports to match. Don't change any other files yet — I want to review the rename before propagating to callers."

Compare to a vague prompt:

> "Rename processCharge."

The vague prompt produces an unpredictable diff. Maybe it changes only one file. Maybe it changes twenty. Maybe it picks a different new name. The specific prompt produces a diff you can predict before you read it, which makes the review fast.

The Diff Review Interface

When Manual proposes changes, the editor opens a diff review panel. The panel shows:

- **A file tree** of all files that will change, with per-file line counts (added/removed)
- **A unified diff view** for the currently-selected file, with hunks separated visually
- **Per-hunk accept/reject buttons** so you can pick which parts of a file to apply
- **A header summary** describing what the model thinks the change accomplishes
- **An apply/discard pair** at the bottom to commit or abandon the whole proposal

The book's habit recommendation: read every line, every time, even when the diff is "obviously correct." The two ways Manual edits go wrong:

- **The model misreads a constraint** and the diff has a subtle behavior change (a default argument changed, an error message slightly altered, an early-return condition flipped). Reading carefully catches these; trusting the model misses them.
- **The model overreaches** and the diff includes "while I was here" cleanup the model thought you'd want. Sometimes the cleanup is good; sometimes it's outside the change you wanted to make and adds review burden you didn't sign up for. Reject the unwanted parts and keep the rest.

A common pattern is a first pass of accept-everything-clearly-right, then a second pass on the parts that need closer reading or follow-up questions. The diff review is interactive — leave parts pending while you ask the model a clarifying question in the same chat.

Partial Accepts and Follow-ups

You can accept some hunks of a file and reject others. After a partial accept, the file has the accepted changes; the rejected hunks are gone (or rather, the original code is preserved). You can then ask Manual to redo the rejected parts with a different approach in a follow-up turn.

The pattern that works well:

1. **First pass:** Accept everything obviously correct. Reject everything obviously wrong.
2. **Second pass:** For anything ambiguous, ask a follow-up. "In the rejected hunk in `auth.ts`, you removed the null check on `session`. Was that intentional? Why?"
3. **Third pass:** Based on the model's explanation, decide whether to re-propose with a constraint ("redo the previous change but keep the null check") or accept the rejection and move on.

This rhythm — first-pass scan, second-pass questions, third-pass refinement — turns Manual from "occasionally produces good diffs" into "reliable workflow."

When Manual Beats Agent

A common question: if Agent can do everything Manual does plus run commands, why use Manual at all?

The answer: review fidelity.

Agent's loop means the model reads its own output (file contents after edits, command output, test results) and continues to act on what it reads. The resulting diffs can be the cumulative output of many sub-actions, which makes them harder to review with full understanding. Manual's single-turn nature produces one focused diff per round, which keeps the review surface small.

Three cases where Manual wins on review:

The change is high-stakes. Anything touching authentication, billing, security boundaries, data migrations, or production configuration. You want a small focused diff you can review carefully, not the cumulative output of an agent loop. Manual gives you that focus.

You have a specific implementation in mind. "Use a Map keyed by user ID, not an array, for the cache" — a specific implementation choice. Agent might or might not pick that approach when given the freedom; Manual implements what you asked for.

You're learning the codebase. When you're new to a project, you want to read every change carefully. Agent's larger output makes this slower. Manual's smaller, focused diffs are paced for careful reading.

The book's heuristic: if the cost of a wrong change is high, Manual. If the cost of slowness is high, Agent. Most working developers find themselves in Manual more often than expected.

Walkthrough: A Multi-File Rename

A concrete walkthrough — the kind of rename the IDE's built-in refactor can't handle because the symbol is referenced via string in a few places.

Scenario: You have a feature flag named `useLegacyCheckout` that you're renaming to `enableNewCheckout` (with the polarity flipped). The flag is referenced in:

- `feature-flags/index.ts` — the flag definition.
- `checkout/router.ts` — the runtime check.
- `tests/checkout.test.ts` — tests for both branches.
- `docs/feature-flags.md` — the docs.
- A few places in `analytics/events.ts` where the flag name appears as a string in event payloads.

The IDE's symbol-based rename won't handle the doc, the test names, or the analytics strings. You could grep and edit by hand, but coordinating five files atomically is Manual's strength.

Step 1: Open chat in Manual mode. Type:

> "Rename the feature flag `useLegacyCheckout` to `enableNewCheckout` across the codebase, flipping the polarity. Sites that previously checked `useLegacyCheckout === true` should now check `enableNewCheckout === false`, and vice versa. Update every reference, including string references in analytics events, test names, and docs. Don't change any logic — just the name and polarity check."

Step 2: Wait for the proposal. A few seconds later, the diff review panel opens with five files. Skim the file tree first — confirm the model identified all five files you expected. If it missed one (say, a sixth file with a string reference), reject the proposal, expand the prompt with the missed file's path, and re-propose.

Step 3: Read each file's diff. For each file, focus on:

- Is the symbol renamed correctly everywhere?
- Are the polarity flips correct? (This is where the model can subtly fail.)
- Is anything else changed you didn't ask for?

Step 4: Accept correct hunks, reject wrong ones. If the polarity flip is wrong in one place, reject that hunk and ask:

> "In `checkout/router.ts`, line 47, the polarity flip looks wrong. The original code routes to legacy when `useLegacyCheckout === true`. The new code should route to legacy when `enableNewCheckout === false`. Re-propose just that line."

Step 5: Apply. Once every hunk is correct, hit apply. The file changes are written.

Step 6: Verify. Run your test suite. If the test names are now `enableNewCheckout: true → routes to new` etc., your tests should still cover both branches and still pass.

The whole walkthrough takes 5–10 minutes. The same task done by hand takes 20–30 and produces more bugs. Manual's win is the combination of model-driven coordination and human-driven review.

Common Manual Mistakes

Three patterns to avoid:

Accepting the diff because "the model usually gets this right." The model usually does. Sometimes it doesn't. The cases where it doesn't are the cases that produce production bugs, and they're often the ones that look correct on a casual read. Read carefully every time.

Reaching for Agent on a Manual task. If the prompt would have been a clean Manual prompt but you reached for Agent because Agent is the impressive mode, you probably burned 3–5x the credits and got back a diff that's harder to review. Reach for the smallest tool that fits.

Not using the partial accept feature. The diff review interface lets you take some changes and reject others. Many users either accept everything or discard everything because they don't notice the per-hunk controls. Use them — they're why the review interface beats "apply patch."

Manual is the focused middle gear. Agent — the third mode — trades that focus for autonomy: the editor takes the wheel and works through multi-step tasks with file edits, terminal commands, and read-evaluate loops. That's Chapter 8.

Chapter 8: Agent Mode — Autonomous Execution

Agent is the headline mode. The model reads files, edits files, runs terminal commands, fetches web pages, reads command output, and iterates – all in a single session, with you watching and intervening when needed. A well-scoped Agent run can take a ticket from "needs implementation" to "PR ready for review" without you typing more than a few clarifying messages. A poorly-scoped Agent run can burn $15 of credits in twenty minutes and produce a diff worse than what you'd have written by hand.

This chapter covers what Agent does, when it's the right tool, how to set it up to succeed, and how to spot it going off the rails before the credit pool drains.

What Agent Does

When you set chat mode to Agent and send a request, the model enters a loop:

1. **Read.** It reads relevant files, runs searches, executes commands to gather information.
2. **Plan.** It decides what to do next based on what it has read.
3. **Act.** It edits files, runs commands, fetches data.
4. **Observe.** It reads the results – file contents, command output, test failures.
5. **Iterate.** Based on observations, it loops back to step 1.

The loop continues until the model decides the task is complete or until you intervene. In practice, Agent runs span anywhere from one minute (small, well-scoped tasks) to thirty-plus minutes (large tasks where the model iterates on test failures).

What Agent can do during the loop:

- **Read any file** in your project that isn't excluded by `.cursorignore`.
- **Edit files** by proposing a diff and applying it. You see each diff as it's applied; you can intervene at any point.

- **Run terminal commands** with your approval. The first time a command runs in a session, you're prompted to allow it; you can grant permission for the rest of the session or a one-time approval.
- **Fetch web pages and documentation** to look up unfamiliar APIs or read external context.
- **Use installed MCP servers** to interact with external systems (GitHub, Linear, Sentry, Postgres, etc. — covered in Chapter 22).
- **Stop and ask you questions** when it needs clarification, hits a permission boundary, or recognizes ambiguity.

What Agent cannot do:

- **Push to the remote** without an explicit step that you authorize.
- **Open pull requests** without your approval.
- **Run privileged commands** (sudo, system-level operations) — those require explicit per-command approval.
- **Continue past a hard error** that prevents the loop from making progress.

When Agent Is the Right Tool

Agent shines in four situations:

End-to-end tickets. A well-scoped feature ticket — "add a `created_at` column to the orders table, update the API to expose it, update the frontend types and components" — is Agent's sweet spot. It writes the migration, runs it locally, updates the backend handler, runs the test suite, fixes any failures, updates the frontend types, and surfaces a coherent PR. The autonomy across layers is the win.

Iterative debugging. "This test is flaky; figure out why and fix it." Agent runs the test repeatedly to reproduce, reads the failure output, hypothesizes, edits, re-runs, and converges. The read-act-observe loop is what makes this work — Manual mode can't iterate on test output the way Agent can.

Refactors that span layers. "Switch this service from REST to gRPC." A change that touches transport, server-side handlers, client-side calls, generated types, and tests is Agent territory. Manual could propose the diff in theory, but a refactor of that scope benefits from Agent's ability to run the test suite at each step and fix forward.

Setup and tooling tasks. "Set up Playwright in this repo, add a basic test for the homepage, configure CI to run it." Agent installs dependencies, creates config files, writes the test, runs it, and commits the working state. Manual would propose the changes; Agent confirms they work.

When Agent Is the Wrong Tool

Agent is the wrong tool more often than it gets credit for. Five cases:

The change is small and specific. A one-file rename, a parameter addition, a docstring update — anything Manual handles in one round. Reaching for Agent here adds the overhead of the agent loop without the benefit. Agent for small tasks is overkill that costs more credits and produces messier diffs.

You don't yet understand what you want. Agent rewards specific instructions. Vague Agent prompts ("clean this up" or "make this better") produce wide-ranging changes that touch more than you wanted, in directions you didn't predict. If you're unsure what you want, switch to Ask first to clarify. Then Manual or Agent for the specified change.

You're working on something high-stakes. Anything touching authentication, billing, security, data integrity, or production configuration. Agent can do the work; the question is whether you want a single focused diff to review (Manual) or a sequence of agent actions whose cumulative effect can drift from what you intended (Agent). For high-stakes work, Manual's review fidelity earns the extra time.

You have a tight credit budget. Agent runs are the expensive part of Cursor. A long Opus 4.7 Agent session can drain your monthly credits in a single afternoon. If your plan has limited credits and you need them to last the month, default to Manual and reach for Agent only when the autonomy is genuinely needed.

Long sessions where context is decaying. Agent quality drops in long sessions as the model's working context fills with file reads, command output, and partial work. The 90-minute Agent run that started well often ends poorly.

Chapter 29 covers this in detail; the short version: commit progress and start fresh sessions before the model loses the thread.

The Scope-Plan-Implement-Polish Workflow

The highest-leverage skill for Agent users is scoping the work before invoking it. The pattern that produces consistent results:

1. Scope. What's the boundary of the work? Which files, which behavior, what outcome? Use Ask mode to clarify if needed. Write the scope down — even just in your head — before sending the Agent prompt.

2. Plan. What are the steps Agent should take? You don't need to micromanage, but a one-paragraph plan ("create the migration, run it, update the handler, update the tests, run the tests") gives the model a structure that's easier to follow than a high-level goal.

3. Implement. Send Agent the prompt. Watch it work. Intervene when it goes wrong.

4. Polish. Once Agent reports completion, read the diff carefully. Run your tests. Run anything Agent didn't run. Make small follow-up corrections by hand or in additional Agent rounds.

Scope and plan are the cheapest two minutes you can spend; implement is where the credits go. The ratio of scope-to-implement work (two minutes thinking, twenty minutes of Agent execution) is what makes the leverage real. Skipping scope is the fastest way to make Agent expensive and frustrating.

Writing Good Agent Prompts

An Agent prompt is different from a Manual prompt. Manual is "make this exact change." Agent is "accomplish this outcome." The prompt should be:

Outcome-focused. "Add a `created_at` column to the orders table and expose it through the API and frontend." Not "edit `migrations/2026_05_orders.sql`, then `api/orders.go`, then `frontend/types/order.ts`..." The first describes the destination; the second describes the path. Agent finds paths better than it follows them.

Constrained on what not to do. "Don't change any unrelated code. Don't introduce new dependencies. Preserve all existing tests." Constraints separate a focused diff from one that includes the model's "while I was here" wandering.

Explicit about success criteria. "The new column should be `TIMESTAMPTZ NOT NULL DEFAULT NOW()`. The test in `__tests__/orders.test.ts` should pass without modification. The API should return the column in ISO 8601 format." Specific success criteria tell Agent when it's done — and let you verify the completion claim.

Concrete about the environment. "Run migrations with `npm run db:migrate`. Run tests with `npm test`. The dev database is already running locally." Telling Agent which commands to use saves it from guessing — and saves you the credits Agent would burn trying three wrong commands.

A complete Agent prompt:

> "Add a `created_at` column to the orders table.
>
> - Migration: `TIMESTAMPTZ NOT NULL DEFAULT NOW()`. Run with `npm run db:migrate`.
> - API: expose `createdAt` in the orders response, formatted as ISO 8601. The handler is in `api/orders.go`.
> - Frontend: add `createdAt: string` to the Order type in `frontend/types/order.ts`.
> - Tests: add coverage for the new field in both the API tests (`api/orders_test.go`) and the frontend (`frontend/components/__tests__/Order.test.tsx`). Run with `npm test`.
> - Don't change any unrelated code. Don't introduce new dependencies.
>
> When the migration has run, all tests pass, and the field is plumbed through, summarize what you did and stop."

The prompt is verbose, but it produces a focused agent run with predictable scope. The verbosity buys reliability.

Watching Agent Work

Once you send an Agent prompt, the chat panel shows a live log of what Agent is doing – file reads, file edits, command executions, command outputs. Watch the first minute carefully and check in periodically thereafter.

Signs Agent is going well:

- It reads relevant files first, then plans, then edits.
- The edits are small and focused per step.
- It runs the tests after substantial changes.
- It reports clear progress.

Signs Agent is going off the rails:

- It edits files outside the scope you set.
- It runs commands you didn't expect (installing packages, modifying configuration).
- It keeps making the same kind of edit and re-running the same test that keeps failing.
- It writes increasingly complicated code to work around an issue that suggests the approach is wrong.
- It has been running for fifteen minutes without visible progress.

When you see those signs, intervene. You can:

- **Pause Agent.** Click pause in the chat panel. Agent stops at its current step.
- **Send a course correction.** "Stop and reconsider. The test failure suggests your migration is wrong, not the test. Fix the migration." Agent reads the message and adjusts.
- **Stop and switch modes.** If Agent is genuinely stuck or producing bad output, stop the run, switch to Ask or Manual, and take over.

The book's habit recommendation: don't let an Agent run go more than five minutes without checking in. The credit cost of an unsupervised Agent run that goes wrong is real money.

Catching Agent Going Off the Rails

The most common Agent failure modes and what to do about each:

Scope creep. Agent decides "while I'm here, I'll fix this other thing." The fix may be reasonable, but it's not what you asked for. Reject the unwanted edits in diff review (Agent still produces reviewable diffs at the end of each step) and clarify the scope in a follow-up: "Don't make changes outside the orders feature."

Workaround spirals. Agent hits a test failure, makes a change to "fix" it, the test still fails, Agent makes another change, and so on. By round three, the code is an unrecognizable mess of workarounds. Stop the run. The real problem is usually upstream — wrong assumption about an API, misread requirement, broken test fixture. Switch to Ask, find the root cause, restart Agent with the corrected understanding.

Confidence drift. Agent claims completion but the code doesn't work. The model said "all tests passing"; you run the tests yourself and three fail. Rare but real, especially in long sessiòns. Always verify Agent's "done" claim before merging anything.

Credit blowups. Agent runs longer than expected. You wake up to a notification that the credit pool is at 80%. Pause, review what's been done, and decide whether to continue (with tighter scope) or take over manually.

Permission boundary fights. Agent wants to run a command you don't want to grant. Sometimes the right answer is granting the permission with care; sometimes it's stopping Agent and doing the privileged step yourself. Don't auto-approve everything Agent asks for — the per-command prompt is a feature, not friction.

Walkthrough: A Real Ticket End-to-End

Concrete example. Ticket: "Add support for product search by SKU."

Step 1: Scope in Ask.

> "Walk me through how product search currently works in this codebase. What's the search interface, where's it implemented, and where do queries get resolved?"

Read the answer. Confirm your understanding.

> "If we were going to add SKU search to the existing product search, what would the change look like? Outline the work."

Get a plan. Push back where it's wrong.

Step 2: Switch to Agent. Send a scoped prompt.

> "Implement SKU search for products.
>
> - The search interface is `searchProducts(query, options)` in `services/search.ts`. Add an optional `sku?: string` to the options.
> - When `sku` is present, filter results to products with matching SKU. The match should be exact, case-insensitive.
> - When `sku` is absent, behavior should be unchanged.
> - Update the API handler in `api/products/search.ts` to pass through the new option from the query string.
> - Add tests: SKU match, SKU no match, SKU absent (current behavior preserved). Run with `npm test`.
> - Don't change any unrelated code. Don't change the search algorithm itself.
>
> When tests pass, summarize and stop."

Step 3: Watch. Agent reads `searchProducts`, the API handler, the existing tests. Edits each in turn. Runs the test suite. One test fails — Agent reads the failure, sees the test fixture didn't have a SKU on the test product, edits the fixture, re-runs. Tests pass. Agent stops and summarizes.

Step 4: Polish. Read the full diff. The code looks right. Run the tests yourself to verify. Check test names — Agent named them sensibly. Run a manual sanity check by sending a query through the dev server. The feature works.

Step 5: Commit. Commit the changes, push, open the PR. Total time: 15–20 minutes. The same ticket implemented by hand might take 45–60. The Agent leverage is real, but it depended on the scoping in Step 1.

Common Agent Mistakes

Three patterns to avoid:

Sending vague prompts and hoping. "Make the search better" is not an Agent prompt; it's an invitation to spend $5 of credits on whatever the model thinks "better" means today. Specific outcomes, specific constraints.

Walking away during the run. Background Agent (covered in Chapter 16) is for "kick off and walk away" workflows. Foreground Agent is interactive. Walking away from a foreground Agent run that needs your input or is drifting off-scope is how credit blowups happen.

Auto-accepting Agent's diffs. Diff review is your last chance to catch issues. Read every line. The model is right most of the time and wrong sometimes; the review pass is your protection against the wrong cases.

Chapter 9: Choosing Between Ask, Manual, and Agent

You've seen all three modes. The question this chapter answers comes up dozens of times a day in actual use: which mode for this task, right now? Get it right and the work flows. Get it wrong and you burn credits on Agent for what should have been Manual, fight Manual on what should have been Agent, or stall in Ask when you should already be implementing.

The chapter is short on theory and heavy on the rubric. The rubric is what you internalize until mode choice becomes automatic.

The Three-Question Rubric

For any task that's larger than a Tab completion or a Cmd+K edit, ask three questions in this order:

1. Do I understand what I want to change?

If no — Ask mode. Don't proceed until the answer is yes. Implementing a change you don't understand is the most reliable way to introduce bugs.

If yes, continue.

2. Does the change require running things to verify?

If yes — Agent mode. The read-act-observe loop is what you need. Manual proposes; Agent verifies.

If no — Manual mode. Single-turn proposal-and-review is faster and cheaper.

3. (Implicit) Is the cost of a wrong change high?

If yes, even if Agent would technically work — bias toward Manual. A smaller, more focused diff is easier to review carefully. High-stakes work deserves high-fidelity review.

That's the rubric. Three questions, under ten seconds to walk through, and it classifies the vast majority of tasks correctly. Internalize it and mode-switching becomes automatic.

A Decision Table

For when the rubric is too abstract, here's a concrete mapping by task type:

Task	Mode	Why
One-line change while typing	Tab	Smallest tool fits
Single-block edit (rename, refactor a function)	Cmd+K	Scoped inline edit
"Explain this codebase"	Ask	Read-only synthesis
Drafting a PR description	Ask	Output is text, not code
Multi-file rename when symbol-rename refactor doesn't catch it	Manual	Coordinated proposal, focused review
Adding an optional parameter through a module	Manual	Clear scope, no execution needed
End-to-end ticket implementation	Agent	Spans layers, needs verification
"Why is this test flaky" with iterative fixing	Agent	Read-act-observe loop is the point
Setup/tooling task (install a library, wire it in, add a smoke test)	Agent	Multi-step, needs execution
"Should I use approach A or B" exploration	Ask	Decision-making, not implementation
Drafting a design doc	Ask	Synthesis, not change
High-stakes change to authentication	Manual	Review fidelity over speed
"Migrate from REST to gRPC"	Agent	Cross-layer, iterative
"Standardize logging across this module"	Manual	Coordinated mechanical change

The table is a starting point. With practice you stop needing it; the rubric runs in your head.

Mode-Switching Mid-Task

Mode isn't a permanent setting for a session. The chat panel preserves history across switches, which means a single task can move fluently through all three modes. The pattern is common and worth practicing:

Phase 1: Ask. Build the mental model. Validate the plan.

> "Walk me through how the rate limiter works. Where is it configured, and what backs it?"

Two or three rounds of Ask until you have the picture.

Phase 2: Switch to Manual. Implement a focused change.

> "Add a per-user rate limit configuration to the rate limiter, defaulting to the existing global limit. Don't change the storage layer yet."

Manual proposes a diff. You review and accept.

Phase 3: Switch to Agent. Execute the broader follow-up.

> "Now wire the new per-user limit through the API and frontend, and add tests. Run them and make sure they pass."

Agent does the layered work that benefits from execution.

Phase 4: Switch back to Ask. Investigate something that came up during implementation.

> "The test for the per-user limit failed initially because the storage layer doesn't have a per-key TTL. Is that something we'd want to add as a follow-up, or is the global TTL fine?"

Ask answers. You decide. If a follow-up change is needed, switch back to Manual or Agent and continue.

This rhythm — Ask to understand, Manual or Agent to implement, Ask when something surprises you, back to Manual or Agent — is the daily texture of fluent Cursor use. Switching friction is essentially zero (one click in the mode dropdown), so the cost of switching often is low. Use that.

The "Scope, Plan, Implement, Polish" Daily Habit

Tying the modes back to the workflow discipline introduced in earlier chapters: the four-phase pattern that consistently produces clean results is:

Scope (Ask). What is the boundary of this task? Which files, which behavior?

Plan (Ask). What are the steps? In what order?

Implement (Manual or Agent). Execute the planned work. Manual for focused changes, Agent for layered or iterative work.

Polish (Manual or by hand). Read every diff. Run every test. Make small follow-up corrections.

For a typical mid-sized task, the time split looks like:

- Scope: 2–5 minutes
- Plan: 2–5 minutes
- Implement: 10–30 minutes
- Polish: 5–10 minutes

Scope-and-plan time feels like overhead compared to "just start coding." It isn't. Implementation runs cleaner because the model knows where it's going and you know what to expect. Polish is faster because the diff is what you predicted. Total elapsed time on the task drops, and result quality rises.

Common Failure Modes — and Recovery

Failure: You started in Agent and you're losing the thread. Agent's been running for fifteen minutes, the diff is sprawling, and you've lost track of what's actually changing. Stop. Switch to Manual. Read the current state of the diff. Decide whether to keep it (commit, then start fresh) or revert (`git restore` the changes Agent made, then restart with a tighter scope).

Failure: You started in Manual but the change keeps growing. What looked like a focused refactor turned out to need migration changes, test fixture updates, and a config tweak you didn't anticipate. Don't keep growing the Manual prompt. Either: (a) split into separate Manual rounds, one per concern; or (b) switch to Agent now that the scope has expanded into territory where execution helps.

Failure: You're in Ask and you're stalling. You've asked the same question three different ways. The model is giving you reasonable answers. You haven't started implementing. Switch to Manual or Agent and start. If you discover a real gap in your understanding, switch back to Ask then — but don't pre-emptively over-explore.

Failure: You're in Manual but the model can't propose a coherent diff because it doesn't know enough. The diff comes back vague or wrong because the model lacks the context to produce a concrete change. Switch to Ask, attach more @ references, build the context, then switch back.

Failure: You're in Agent but the loop keeps failing on the same step. Agent's been re-trying the same fix for three rounds and the test is still failing. Stop. Read the failure yourself. The root cause is likely upstream of where Agent is poking — different test fixture, different assumption, different file. Switch to Ask, find the root cause, restart Agent with the right understanding.

The pattern across all failures: notice quickly, switch modes deliberately, don't keep doing what isn't working. Cursor's mode flexibility is the recovery path; using one mode out of habit is what makes failures expensive.

A Day in Three Modes

To make this concrete, here's roughly what a productive day in Cursor looks like:

9:15 AM. Open the project. Read the morning's GitHub notifications. A teammate left a question on yesterday's PR: "Why this approach instead of X?" Switch to Ask, attach the PR's diff, ask the model to draft a response explaining the tradeoff. Edit the draft. Post the comment.

9:45 AM. First substantive task: a small bug fix from the backlog. The fix touches one file. Cmd+K with the function selected, prompt for the fix, accept the diff. Run the test. Commit. PR.

10:30 AM. Bigger task: add a new admin-only API endpoint. Switch to Ask. Walk through how existing admin endpoints work. Draft a plan. Switch to

Manual. Implement the endpoint. Review the diff. Switch to Agent for test coverage and the integration test. Watch. Verify. Commit. PR.

12:00 PM. Lunch.

1:00 PM. Investigate a flaky test reported overnight. Switch to Agent. "This test is flaky; figure out why." Watch Agent reproduce, diagnose, fix. Verify. Commit. PR.

2:30 PM. Refactor task: rename a confusing variable across the codebase. Switch to Manual. One round, focused diff. Review. Apply. Tests still pass. Commit.

3:30 PM. Meeting.

4:30 PM. Review a colleague's PR. Switch to Ask in their branch. "Walk me through what this PR does." Read their changes alongside the model's summary. Leave thoughtful comments.

5:30 PM. Last task of the day: write release notes for the week's commits. Switch to Ask. "Draft release notes covering these last twenty commits." Edit. Post.

6:00 PM. Done.

In a day like that, you've probably switched modes thirty times. None felt like friction because each fit the work. That's the goal — not "use Agent for everything" or "use Manual for everything" but mode fluency that makes the right tool the obvious choice.

A Habit, Not a Rule

The rubric is a starting habit, not a rigid rule. With a few weeks of practice, you'll start to know — without consciously running through the three questions — which mode the next task wants. That intuition is the goal. Until you have it, run the rubric explicitly. The forced reflection is what builds the intuition.

From Modes to Inline

Part II is done. You know the three modes, when to use which, and how to switch mid-task. Part III shifts to the inline-editing primitives that quietly determine whether the editor feels like leverage or like work — Tab autocomplete, Cmd+K, and the discipline of reading diffs critically. Chapter 10 starts with Tab: the small, fast model that runs constantly in the background and, used well, becomes nearly invisible.

Part III: Tab and Inline Edits

"The fastest tool is the one that disappears into the work."

— A working principle

Chapter 10: Tab Autocomplete

Tab is the feature you'll use most. By a wide margin. By the end of a typical workday, an active Cursor user has accepted hundreds of Tab proposals, ignored thousands, and probably never opened the chat sidebar more than a dozen times. The chat features get the demos and the headlines; Tab gets the actual hours.

That asymmetry is the whole reason this chapter exists separate from the chat-mode chapters. Getting Tab right — accepting the right ones, ignoring the wrong ones, and tuning your behavior around what Tab is and isn't good at — is where most of Cursor's daily leverage comes from. This chapter is about getting that right.

What Tab Is

Tab is Cursor's name for the inline ghost-text completions that appear as you type. The completion model is small, fast, and tuned by Cursor specifically for codebase-aware completion. It's not the same model that powers chat — those are the frontier models from Anthropic, OpenAI, DeepSeek, and Google. Tab uses a separately-trained model that prioritizes latency (most completions arrive in well under a second), context awareness (it sees substantial parts of your codebase, not just the current file), and the specific shape of "predict what the developer will type next."

When you accept a completion (Tab key), the proposed text commits at the cursor. When you keep typing, the completion either updates to incorporate your new typing or disappears. The interaction is silent and non-blocking — Tab never interrupts your flow with a modal or a confirmation dialog. The cognitive cost is whatever you pay to read the suggestion before deciding whether to accept.

What Tab Is Good At

Four categories of completion where Tab is consistently strong:

Boilerplate and pattern repetition. When you write the second of three similar `if` branches, Tab predicts the third with high accuracy. When you write the first field of a TypeScript interface, Tab completes the rest based on what the interface is for and what nearby interfaces look like. When you've typed `for (let i = 0; i <`, Tab generally knows what you want to iterate over. Anywhere your project has consistent patterns, Tab extends them fluently.

Codebase-aware function calls. The single feature that distinguishes Cursor's Tab from competitors is codebase awareness. When you start typing a function call, Tab fills in arguments that match how the function is used elsewhere in your project. When you write a test, Tab proposes assertions that match the project's testing conventions. When you import a symbol, Tab proposes the import path the rest of your codebase uses, not a generic guess.

Multi-line completions across structural boundaries. Tab can propose a function body, a switch statement, an interface definition – completions that span multiple lines and respect the structure of the language. Multi-line is enabled by default and is one of the things that makes Tab feel meaningfully ahead of older completion tools.

Smart rewrites. Tab can propose edits to lines you've already written when the change is "obvious." Rename a variable in one place and reference it three lines down, and Tab proposes updating the reference. Change a function signature, and Tab proposes updating the call site below. These rewrites appear as multi-line proposals; you accept the same way (Tab) and they apply across the proposed range.

What Tab Is Bad At

Equally important – and the part that prevents bad accepts:

Anything requiring semantic understanding it doesn't have. Tab guesses. It guesses well from local patterns and codebase patterns. It does not understand business logic the way a person who's worked on the project for a year does. When you're writing code that depends on subtle business rules ("orders over $1000 require manual review"), Tab proposes something plausible-shaped but potentially wrong on the specifics. Read carefully.

Code in unfamiliar parts of the codebase. If you've just opened a file you've never seen, Tab's confidence is calibrated to the codebase's general patterns, not your specific intent in that file. It may propose completions that match the project's style but don't match what you're trying to accomplish. The cure: do a few keystrokes manually before relying on Tab in unfamiliar territory.

Cross-cutting concerns that depend on what you're about to do. Tab predicts the next few characters based on what you've typed so far. If you're about to make a structural decision (Promise.all or sequential awaits, throw or return an error), Tab guesses based on local patterns — but the right answer depends on what you're optimizing for. Don't let Tab make those decisions for you by accepting a confident-looking completion without reading.

Anything where being subtly wrong is expensive. Authentication checks, security boundaries, error handling that affects user-facing behavior, data validation rules. Tab proposes plausible code in these areas; the cost of accepting a subtly-wrong completion in security-critical code dwarfs the time saved. Use Tab freely in test code and tooling; read more carefully in production-critical paths.

The Accept-Reject Discipline

The highest-leverage habit for Tab is reading every proposed completion before accepting. This is harder than it sounds.

Tab is fast. Reading is slow. There's a constant temptation to bang Tab when the completion looks "about right" and move on. Hundreds of times a day, that temptation is mostly fine — the completion really is right, and reading it carefully every time would slow you down. But the few times the completion is subtly wrong, the consequences compound. A field renamed to a slightly-different name. A boundary check inverted. An off-by-one in a loop bound. Catching the bug in code review is cheap; catching it in production is expensive.

The discipline that works for most professional users:

Always glance at the completion before accepting. Three-quarters of a second of attention catches the obvious-wrong completions. This isn't deep code review — it's a quick "yes, that matches what I was going to type."

Slow down in critical paths. When you're editing security, billing, data integrity, or anything that runs in production with real consequences, read completions more carefully. Better, type those areas yourself rather than relying on Tab.

Reject completions that are "almost right." A 90%-right completion is not a small win — accepting it and editing the 10% takes longer than typing the whole thing yourself, and the 10% you miss in your edit becomes a bug. Reject and type. Tab will offer a different completion next time as you type more characters.

Notice when Tab is wrong systematically. If Tab keeps proposing completions in a particular file or feature that are wrong in similar ways, the codebase context is misleading the model. Either there's an inconsistency in the project to clean up, or you should add a project rule that overrides the misleading pattern. Chapter 14 covers rules in detail.

When Tab Is Particularly Strong

A few specific situations where Tab earns its keep:

Writing test cases. After you write the first test for a function, Tab reliably proposes the next three or four cases, including edge cases and setup boilerplate. The pattern recognition is good and the surface area for "subtly wrong" is small — tests fail loudly.

Filling in repetitive type definitions. TypeScript interfaces, Go struct fields, Python type hints, schema definitions. Anywhere you're writing a list of typed names that follow a pattern, Tab is fast.

Mechanical code transformations. Converting a list of magic numbers to a constant, applying a renamed variable through a function body, reformatting a destructuring pattern. Local repetition is Tab's wheelhouse.

Boilerplate at structural boundaries. New file headers, the standard imports for the project, the function signature for a method that follows a project convention. Tab gets these right almost every time.

When Tab Is Particularly Weak

Three situations where Tab fails consistently enough that you should default to typing:

Unfamiliar APIs. When you're using a library new to this project, Tab proposes calls based on what it knows about that library — but the project's wrappers, conventions, and helpers aren't in scope until you actually use them once. The first time you call a new dependency, type it yourself; Tab becomes useful by the third call.

Cross-file consistency. Tab sees a lot of your codebase but not all of it. When the consistency requirement is "exact match with this specific other file three directories away," Tab may propose a close-but-not-exact match. For these cases, use Cmd+K with the reference file attached, or Manual mode for explicit coordination.

Code that depends on what you're going to write next. Tab predicts from what you've typed. If the right code depends on what you're about to type after this line, Tab can't see the future. In these cases, write the structural skeleton first (the broad shape of the function), then come back and fill in details — Tab gets stronger as the surrounding context fills in.

Tuning Tab to Your Style

Tab has settings — sensitivity, multi-line aggressiveness, language-specific overrides — but the book's recommendation is to ignore them for the first month. The defaults are tuned to the median Cursor user, and tuning before you have a few weeks of usage under your belt usually makes things worse, not better.

Adjustments worth considering after a month of use:

If Tab is too aggressive in a language you write differently. If Tab keeps proposing multi-line completions in a language where you prefer line-at-a-time control (a configuration DSL, a complex template language, a language where you'd rather think carefully than autocomplete), reduce multi-line aggressiveness for that language in settings.

If Tab is suggesting outdated patterns. If your project recently shifted conventions (new error-handling pattern, new logging library, new testing framework) and Tab proposes the old patterns, the most reliable fix is a project rule that captures the new convention. Tab reads project rules and re-anchors its predictions to them.

If Tab is too noisy in a language with verbose syntax. Some languages produce many low-confidence completions that aren't useful. Adjust Tab's confidence threshold per-language to suppress the marginal proposals.

The general principle: don't tune what you haven't measured. A month of usage data is enough to see patterns; a week isn't.

Tab in Unfamiliar Codebases

Tab's quality scales with how well the model understands your codebase. In a well-indexed project you've worked on for a while, Tab feels nearly clairvoyant. In a project you just opened where indexing is still running, Tab is more often wrong than right.

The implications:

Wait for indexing. When you open a new project, the first ten or twenty minutes of indexing are when Tab is weakest. Don't judge Tab's quality during that window; let the index complete before deciding.

Start with small explicit changes. In a new codebase, the first few edits you make are when Tab has the least to work with. Type more, accept Tab proposals more carefully. After twenty or thirty edits, Tab has a much richer picture and quality climbs.

Use @ references in chat to teach Tab. Tab doesn't read chat history directly, but the codebase index notices files that are frequently accessed. The more you reference files in chat, the more they become "core context" and the better Tab performs in adjacent code. (This is partial — Tab is not a chat-driven feature — but the indexing layer benefits.)

When Tab Disappears

A few situations where Tab won't show suggestions, and what they mean:

You're typing too fast. Tab needs a brief pause to compute. Typing without pauses suppresses the proposal. This is intentional — better to skip a stale completion than show one that's already wrong.

You're in a context where Tab is configured off. Some file types, some markdown blocks, some config files. Settings let you exclude file types from Tab.

The codebase isn't indexed yet. Tab works without the index but is much weaker. If completions are sparse and generic in a project that should be indexed, the index probably isn't ready.

Network or service issue. Tab calls a remote model. If your network is flaky or Cursor's services are degraded, Tab goes quiet. Cursor's status page (cursor.com/status) is the place to check.

The Feedback Loop

Cursor's Tab model improves over time, and your accept/reject behavior is part of the input. Accepting is a positive signal; rejecting (typing something different) is a corrective signal. You don't manage this loop directly — there's no "rate this completion" button — but the model learns from aggregate patterns.

The implication: accepting completions you don't actually want, just to "save time," teaches the model that those completions were good. Don't do that. Reject what's wrong; accept only what's right. Accept rate isn't a metric to maximize — it's a metric of how well Tab is matching your intent. Optimize for the latter.

Tab handles the smallest changes well. The next tool up — Cmd+K — handles the deliberate medium-sized ones, when you want a scoped edit rather than a continuation. Chapter 11 covers it.

Chapter 11: Cmd+K Inline Generation

Cmd+K is the second-most-used AI feature in Cursor, after Tab. It's what you reach for when Tab isn't enough — when you want a specific edit rather than a continuation, when you want to ask for something rather than have it predicted, when you want to write a new function rather than autocomplete one. The keystroke (Cmd+K on macOS, Ctrl+K on Windows and Linux) opens an inline prompt at your cursor or selection; you type a request, and a proposed change appears in place.

Cmd+K is the workflow Cursor users talk about least. Tab is famous; chat is the demo; Cmd+K is the quiet workhorse. This chapter is about getting it right — when to use it, what makes a good prompt, and the patterns that turn it into a faster path than the alternatives.

What Cmd+K Does

When you press Cmd+K, an inline prompt box appears at your cursor location or — if you have text selected — over the selection. You type a natural-language description of what you want, and the model proposes a change.

Behavior differs slightly based on whether anything is selected:

With no selection. Cmd+K is a "generate at this location" prompt. The model writes new code at the cursor. Use this for "write a function that does X" or "add a comment block here describing Y."

With a selection. Cmd+K is a "rewrite this selection" prompt. The model proposes a replacement for the selected text. Use this for "refactor this," "convert this to TypeScript," "add error handling to this."

In both cases, the proposed change appears as an inline diff — additions in green, deletions in red, with the prompt box still visible above. You accept with Cmd+Enter (or click Accept), reject with Esc, or refine by typing a follow-up prompt to iterate.

The whole interaction stays inline. No sidebar, no chat panel, no diff review tree. The change appears where your cursor was, you decide, and the editor returns to its normal state. Latency is typically a couple of seconds for small edits, longer for larger generations.

When Cmd+K Beats Tab

The line between Tab and Cmd+K is intent.

Tab is predictive — it guesses what you're about to type next. The right Tab is the one that matches what you would have typed.

Cmd+K is intentional — it executes a specific request. The right Cmd+K is the one where you tell the model what you want and it produces it.

Reach for Cmd+K instead of Tab when:

You know what you want and Tab doesn't. Tab is guessing from context; if your intent isn't well-predicted by your recent typing, Tab proposes something else. Cmd+K lets you state the intent directly. "Write a function that takes an array of orders and groups them by customer, returning a Map" is a Cmd+K prompt; you'd never get that from Tab without typing most of it.

The change is bigger than a few characters and you don't want to scaffold it manually. A new utility function, a new test case, a new component file — too large for Tab, too small for chat. Cmd+K hits the size right.

You want to rewrite existing code rather than continue from it. Tab continues; Cmd+K rewrites. To convert a class component to a function component, refactor a function for clarity, or add error handling to existing code, Cmd+K with a selection is the cleanest path.

You want a one-shot result, not a conversation. Cmd+K is single-prompt. You ask, you get a proposal, you accept or reject. Chat is conversational. For changes that don't need back-and-forth, Cmd+K is faster and lighter.

When Cmd+K Beats Chat

The line between Cmd+K and chat is conversation depth.

Chat is for tasks that benefit from back-and-forth — exploring, planning, iterating across multiple steps. Cmd+K is for tasks where one prompt is enough and the result lives in a specific place.

Reach for Cmd+K instead of chat when:

The change is local. Inline edits at a specific location belong to Cmd+K. The chat panel adds friction (open chat, attach context, prompt, review, apply) that's wasted when the change is "rewrite this selection."

You don't need to explain the broader context. Chat is good for tasks where you spend time framing the problem. Cmd+K assumes you know what you want; the inline prompt is a one-line way to express it.

You're not looking for a discussion. Sometimes you want a thinking partner — chat. Sometimes you just want the change made — Cmd+K.

The result is a single localized edit. If the answer is "edit these three lines," Cmd+K. If the answer is "edit five files coordinated together," chat with Manual or Agent.

When Cmd+K Is the Wrong Tool

Three cases where Cmd+K is not the right reach:

Multi-file changes. Cmd+K is single-file. The selection is your scope. A coordinated change across multiple files is Manual or Agent territory in chat. Pushing Cmd+K beyond its single-file shape gives you partial answers and missed sites.

Tasks requiring real exploration. "How should I approach refactoring this service?" is a chat task — Ask mode, specifically. Cmd+K is for executing an idea you have, not for figuring out what idea to have.

Anything you want to discuss before applying. Cmd+K is one-shot. If you'd benefit from "let me see your plan first, then implement," that's chat. The lack of conversational depth in Cmd+K is the feature when you don't need it and a limitation when you do.

Writing Good Cmd+K Prompts

The Cmd+K prompt is short by design. The inline box doesn't accommodate long, paragraph-style prompts well. The constraint is intentional — Cmd+K is for asks that fit in one or two sentences. Longer prompts are a tell that you should be in chat.

Prompt patterns that work consistently well:

Verb-led action prompts. "Refactor this to use async/await." "Add type annotations." "Convert to a class component." "Extract the validation logic into a separate function." Lead with the verb, name the change, done.

Specification prompts for new code. "A function that takes a list of users and returns the ones with active subscriptions, sorted by last login descending." Specify inputs, outputs, ordering. The model fills in the implementation.

Constraint-augmented prompts when defaults are wrong. "Refactor this for readability but keep the same time complexity." "Add error handling but don't change the function signature." "Rewrite this in pure functions, no class state." Constraints keep the model from drifting in directions you don't want.

Negative prompts when something keeps going wrong. If the model keeps adding a comment style you don't want, or proposing a library you're not using, tell it not to: "Don't add JSDoc comments. Don't import any new libraries."

A complete Cmd+K prompt:

> "Refactor this to use async/await instead of promise chains. Preserve error handling — every catch should still be there. Don't introduce any new dependencies."

Two sentences, specific, constrained. The diff that comes back is predictable.

Compare to a vague Cmd+K prompt:

> "Make this better."

The vague prompt produces unpredictable changes. The model has to guess what "better" means — readability? performance? error handling? type safety? — and often picks something different from what you wanted.

Common Cmd+K Patterns

Patterns that come up often enough to internalize:

Generate from a comment. Write a comment describing what you want, hit Cmd+K with the comment selected (or with the cursor on the line below it), and prompt: "Implement what the comment describes." The comment serves as both requirement and spec.

Add boilerplate at a function definition. Position your cursor at a function signature. Cmd+K: "Add JSDoc/docstring documenting parameters and return value." The model reads the function body to infer what to write.

Convert between equivalent forms. Selection on a `for` loop, Cmd+K: "Convert to .map." Selection on a switch statement, Cmd+K: "Convert to a lookup table." Selection on chained method calls, Cmd+K: "Rewrite as discrete steps with intermediate variables."

Generate test cases. Selection on a function definition, Cmd+K: "Add tests covering happy path, empty input, and the boundary at zero." The model writes tests inline — accept and move them to your test file, or use Manual mode if you want test files coordinated.

Apply project conventions. Selection on code that doesn't follow your project's patterns, Cmd+K: "Rewrite this to match the project's logging convention." The model uses the codebase index to find the convention and apply it.

Refactor for a stated goal. Selection on messy code, Cmd+K: "Refactor for testability — make dependencies explicit and remove static state." A goal-stated prompt produces more focused refactors than "make this cleaner."

Iterating on a Cmd+K Result

After the model proposes a change, you have a few options:

Accept (Cmd+Enter). The change applies. Done.

Reject (Esc). The proposal disappears, the original code stays.

Type a follow-up in the inline prompt. "That's close, but rename the helper function to something more specific." The model produces a revised proposal. The conversation is short — no chat history beyond the current iteration — but you can iterate two or three times to refine.

Switch to chat. If iteration will require more back-and-forth than the inline prompt accommodates, click "Open in chat" (or escalate manually) and continue in the chat panel with full history.

The follow-up loop is faster than re-prompting from scratch. If the first proposal is 80% right, a follow-up is usually all you need. If the first proposal is wildly off, reject and start over with a better prompt — chasing a bad proposal with iterative tweaks usually takes longer than restarting.

Walkthrough: Writing a Function from a Comment

Concrete example. You're working on a project and you need a utility function. You know what it should do but don't want to type it out.

Step 1: Write a comment describing the function. Type:

```
// Group orders by customer ID and return a Map. Skip orders with no customer.
```

Step 2: Position your cursor on the line below the comment. Press Cmd+K.

Step 3: Prompt. Type:

> "Implement the function described in the comment above. Use TypeScript, generic types where appropriate, and match the project's style."

Step 4: Review the proposal. The model writes a function. Read it. Check that:

- The function name matches your project's naming convention.
- The types are reasonable (the parameter is typed, the return type is explicit).
- The "skip orders with no customer" requirement is implemented.
- The implementation isn't doing anything surprising.

Step 5: Accept or iterate. If it's right, Cmd+Enter. If it's close but the function name should differ, type a follow-up: "Rename to `groupOrdersByCustomer`." Apply.

Step 6: Verify. Use the function somewhere. Tab autocompletes calls now that the function exists in the file. If it works as expected, you're done.

The whole walkthrough takes 30–60 seconds for a function that would have taken 3–5 minutes to type from scratch. Repeat dozens of times a day and the leverage compounds.

Cmd+K and Selection Boundaries

A subtle thing about Cmd+K with selections: the selection boundary defines the editable region, but the model also reads context outside the selection — surrounding code, file imports, related symbols — to inform the rewrite. Usually that's what you want; occasionally it isn't.

If you want the model to consider only the selection (because surrounding code is misleading or irrelevant), say so in the prompt: "Rewrite the selection. Don't reference any code outside the selection." The model will be more literal in its rewrite.

Conversely, if the model is being too literal and missing relevant project context, expand the selection or rely on the model's automatic context. Most of the time, the defaults are right.

Cmd+K Anti-Patterns

Three Cmd+K anti-patterns to avoid:

Long, paragraph-style prompts. If your prompt has more than three sentences or describes multiple concerns, you're using the wrong tool. Open chat instead. The cramped inline prompt is a forcing function for short, focused asks; fighting that constraint slows you down.

Cmd+K for tasks that span files. Cmd+K is single-file. If your prompt mentions other files, your task is in chat territory. Pushing Cmd+K to handle multi-file tasks produces partial results you have to finish by hand.

Cmd+K instead of Tab for line-at-a-time work. If you're using Cmd+K for one-line edits because it feels more deliberate than Tab, you're paying a latency cost (a few seconds per edit) that compounds. Use Tab for the small stuff. Save Cmd+K for the deliberate medium-sized edits.

Chapter 12: Working with Selections and Diffs

The two skills that separate effective Cursor users from frustrated ones aren't about the AI features themselves — they're about how you frame the inputs and how you read the outputs. This chapter covers both: the discipline of selecting before prompting, and the discipline of reading proposed diffs critically.

Not glamorous skills. No flashy feature to demo. But they're the difference between a workflow where the editor feels like leverage and one where it feels like a slot machine.

Why Selections Matter

Model quality depends on knowing what you're asking about. When you invoke Cmd+K or chat in Manual mode, the editor sends the model some context — your prompt, the current file, the codebase index, sometimes other files based on heuristics. Of all that context, the most reliable signal of intent is the selection.

A selection tells the model: "this is what you're working on." With no selection, the model infers scope from the prompt and from where your cursor is. With a selection, the scope is unambiguous.

The implications:

Select before you prompt, when you know the boundary. To refactor a specific function, select the function, then Cmd+K. To rewrite a specific code block, select the block first. The selection does work that would otherwise have to come out of the prompt's wording.

Selections shrink the surface area for misunderstanding. With a selection, the model can't drift into "while I'm here, let me also fix this nearby thing" because the scope is bounded. With no selection, the model is likelier to propose changes outside what you asked for.

Selections improve diff quality. Rewrites of selections produce focused, reviewable diffs. Cmd+K with no selection on a vague prompt produces sprawling output that's harder to review and harder to reject piecemeal.

The general rule: when in doubt, select. The cost of selecting (click-and-drag, or Shift-click) is trivial. The benefit in clearer model behavior is real.

Selection Patterns That Work Well

A few selection idioms that come up often:

Select the whole function for refactors. Place your cursor anywhere in a function and use Ctrl-Shift-Up/Down (or your editor's "expand selection" shortcut) to grow the selection up to the function's boundary. Cmd+K on the whole function: "refactor for clarity." The boundary is exactly the function.

Select a logical block, not a syntactic boundary. To rewrite the validation portion of a function, select the validation block, not the whole function. The model rewrites the selection and leaves the rest alone. The boundary you choose is the boundary the model respects.

Select multiple files in the explorer for chat. When you want chat to consider several files together, select them in the file explorer (Cmd-click each), right-click, "Add to Chat." This is more reliable than @ references for scoping multi-file Manual or Agent tasks.

Select tests when you want them rewritten. To update a test to cover a new case, selecting the existing test gives the model the testing style as immediate context. The new test matches the project's pattern.

Don't select code you want left alone. Reverse of the above. If your selection includes code that should remain unchanged, the model may rewrite it. Select tightly.

When Not to Select

Selecting isn't always the right move:

For "generate something new at this location" prompts, no selection is correct. Place the cursor where the new code goes, Cmd+K with no selection, and the model writes new code at the cursor.

For chat exploration and questions, selection is optional. "How does authentication work in this project" doesn't need a selection — the model uses the codebase index.

For multi-file tasks, the per-selection model breaks down. Use Manual or Agent in chat with @ references and let the model work across files.

For tasks where you want the model to consider broader context, a tight selection can be limiting. "Refactor this function with the rest of the file in mind" is better in chat than a selection-bounded Cmd+K.

Selection is a tool, not a rule. Use it when it helps, skip it when it doesn't.

Reading Diffs Critically

The other half of the chapter: reading what the model proposes. This is the skill most users underinvest in, and it's where the highest-impact bugs creep in.

A proposed diff — from Tab, Cmd+K, Manual, or Agent — has three audiences:

1. **The model**, which wrote it and thinks it's correct.
2. **You**, who has to decide whether to apply it.
3. **The compiler / test suite / production users**, who will discover anything you both missed.

Audiences 1 and 3 are non-negotiable. The model writes what it writes; production discovers what it discovers. Audience 2 — you — is the only point in the loop where intent and result can be reconciled before things go wrong.

What to Look For When Reading a Diff

A reading checklist that catches most issues:

Does the diff do what you asked for? The obvious check. If you asked for a rename and the diff is doing more than renaming, something's off.

Does the diff do anything you didn't ask for? Less obvious and more important. Models occasionally bundle in "while I was here" changes — formatting, restructuring, "minor cleanup" — that weren't in scope. Sometimes these are good; often they add review burden you didn't sign up for. Reject the unwanted parts.

Are the boundaries right? Look at the first line of the diff and the last. Did the model start the change where you expected and end it where you expected? Off-by-one errors at boundaries are common.

Are the conditions right? Wherever there's an `if`, a comparison, a polarity check — read it carefully. The model occasionally inverts a condition or changes a comparison subtly. These are the bugs that compile cleanly and fail in production.

Are the error paths preserved? When refactoring, the success path usually looks right. Error paths get less attention from the model and need more from you. Check that catch blocks still catch what they should, that error messages are preserved if behavior depends on them, that early-return conditions still hold.

Are the names right? The model occasionally renames a variable to something that "reads better" but breaks a convention or shadows a name elsewhere. Read names with as much attention as logic.

Did the model add new dependencies or imports? Check the imports. A new library import you didn't authorize is a flag. The library may be fine; you should know it's there.

Are there any TODO or placeholder markers? Models occasionally leave `// TODO: implement this` markers in incomplete sections. They're easy to miss in a long diff and embarrassing to discover later.

For a small diff (a Tab or Cmd+K result), the checklist takes maybe ten seconds. For a Manual diff covering several files, a minute or two per file. The time is well-spent — these are the catches that prevent the production-bug variant of "but the AI wrote it."

Partial Accepts

Manual mode (and to a lesser extent Cmd+K) lets you accept some hunks of a diff and reject others. Most users underuse this.

When the diff has some hunks you want and some you don't:

Accept the good ones. Click accept on the hunks that are correct.

Reject the bad ones. Click reject on the hunks that are wrong or out-of-scope. The original code stays for those.

Ask a follow-up to redo the rejected parts. "In `auth.ts` you removed the null check on session. Re-propose just that hunk but keep the null check."

Partial-accept lets you take 80% of a useful proposal without the wrong 20%. The alternative — reject everything because some of it is wrong — wastes the 80% that was good.

Some users never notice the per-hunk controls and accept-or-reject the whole diff as a unit. Don't be one of them. The granularity is there for a reason.

The "Review Every Line" Habit

The discipline that separates senior Cursor users from junior ones: every line, always.

This applies to Tab proposals (read each one before accepting), Cmd+K results (read the diff before Cmd+Enter), Manual proposals (read every hunk in every file), and Agent's cumulative output (review the final state before committing).

The habit feels slow at first. Reading takes time. Especially for Tab, where reading every proposal is a constant tax on the speed advantage Tab gives you. The pull to "trust the model and move on" is constant.

Don't trust the model. Verify it. The model is right most of the time and wrong some of the time, and the wrong cases are exactly the ones that look right at a glance.

The good news: the habit gets faster with practice. After a few weeks of reading every proposal, the read becomes part of the typing rhythm. You're not stopping to read; you're reading in the same flow as typing. The cost drops to near-zero, while the benefit — catching the subtle-wrong cases — stays high.

Reading Diffs in Agent Mode

Agent's diffs differ from Manual's. Manual produces one focused diff per round. Agent produces a sequence of small diffs, plus command outputs, plus file reads, accumulated across the loop.

Reading discipline for Agent:

Read each step's diff as it happens, not just the end. Watching Agent work is part of the workflow. When you see Agent edit a file, glance at the diff before it moves on. Catching a wrong direction at step 3 is far cheaper than catching it at step 12.

Pay attention to commands as well as edits. Agent's terminal commands are part of the change. A `rm -rf` in the wrong directory is a worse outcome than a wrong file edit. The first time Agent runs a command, the permission prompt is your last chance to catch a problem.

Read the final state, not just the diff. Agent's cumulative changes can be coherent step-by-step but produce a final state that's confused — leftover scaffolding, unused imports, half-finished refactors. After Agent claims completion, look at the resulting files in their final form, not just the diff against the start.

Run the tests yourself. Agent often runs the tests in its loop. Run them again yourself before committing. "Tests pass" is a claim; a verified pass is the truth.

A Real Case: The Inverted Polarity Bug

A concrete example of the kind of bug that careful diff reading catches.

The task: rename a feature flag and flip its polarity (introduced in the Chapter 7 walkthrough). The Manual diff comes back. Most of it looks right.

In one file, the model produced:

```
// before
if (useLegacyCheckout) {
  return legacyHandler();
}
return newHandler();

// after (proposed)
if (enableNewCheckout) {
  return legacyHandler();
}
return newHandler();
```

A user accepting this without reading carefully ships the bug: the variable was renamed but the polarity wasn't flipped. Both branches now go to the wrong handler.

A user reading the diff carefully catches it: "I renamed and asked for polarity to flip. The if-branch should call `newHandler()` when `enableNewCheckout` is true." Reject the hunk. Re-prompt for the correct flip. Apply.

The bug is small, easy to introduce, and easy to catch — but only with the reading habit. Without it, ship.

Part III closes here. Tab, Cmd+K, the selection-before-prompt habit, and the read-every-line habit are the daily inline mechanics. The leverage that compounds across a project's lifetime — the kind that turns Cursor from "fast editor" into "team-wide multiplier" — lives in features that don't show up in any single inline edit: codebase indexing, project rules, `.cursorignore`, Background Agents, multitask. Part IV works at that scale, starting in Chapter 13 with the index.

Part IV: Working at Scale

"Leverage is what compounds while you sleep. Pay it the attention it deserves."

— A working principle

Chapter 13: Codebase Indexing and Context Retrieval

Every "the model already knew that" moment in Cursor traces back to the index. When Tab proposes a completion that matches your project's conventions, when Ask answers a cross-file question without you attaching the relevant files, when Manual produces a coordinated multi-file diff — that's the codebase index doing the work behind the curtain.

The index is invisible when it's healthy and frustrating when it isn't. This chapter covers what the index is, what gets indexed, what doesn't, and what to do when the index is the bottleneck.

What the Index Is

When you open a project for the first time, Cursor builds a representation of the codebase that the AI features can use as context. The high-level shape:

A file tree. Cursor enumerates the files in the project, respecting `.gitignore` and `.cursorignore`. The list is the universe of files the AI features can reference.

A Merkle tree of file hashes. Each file's content is hashed, and the hashes are organized into a Merkle tree. This lets Cursor detect which files have changed since the last index and update only those, rather than re-indexing everything every time.

Embeddings. Each file (or chunked piece of a file) runs through an embedding model, producing a vector representation of its semantic content. The embeddings live in a vector database (Cursor uses Turbopuffer as of 2026) hosted on Cursor's infrastructure.

Symbol metadata. Function names, class names, type definitions, exports — the structural elements your code defines — are extracted and stored separately from the embeddings, enabling exact-name lookups alongside semantic search.

When you ask a chat question, invoke Manual on a multi-file task, or just type code that Tab is about to complete, Cursor's retrieval layer pulls relevant chunks from the index and feeds them to the model as context. Retrieval is the difference between "AI assistant that sees the current file" and "AI assistant that sees the codebase."

How Indexing Works in Practice

The first time you open a project, Cursor offers to index it. On a small project (a few thousand files, a few hundred thousand lines), indexing completes in 10–30 seconds. On a large monorepo (a million files, hundreds of millions of lines), indexing can take 5–15 minutes — sometimes longer.

Indexing status shows in the bottom-right of the Cursor window: an animated indicator while indexing is in progress, a checkmark when it's complete. You can also see status in Settings → Codebase → Indexing.

After the initial index, Cursor watches for file changes and re-indexes incrementally. A typical save triggers a sub-second re-index of the changed file. Larger changes (a `git pull` that brings in a thousand files) trigger a larger re-index that takes a few seconds. Most of the time, you don't notice incremental indexing happening.

Indexing happens in the background and doesn't block the editor. You can use Tab, Cmd+K, and chat while indexing is in progress — the AI features work with whatever portion of the index has been built so far. Quality climbs as indexing progresses.

What Gets Indexed

The default rules for what's included:

Tracked files in the project. If the file is in version control (or would be — Cursor doesn't strictly require git, but the patterns are similar), it's a candidate.

Files not excluded by `.gitignore` or `.cursorignore`. Cursor respects both. Ignored files are not indexed.

Files of supported languages. Most mainstream languages — TypeScript, JavaScript, Python, Go, Rust, Java, C/C++, Ruby, PHP, C#, Swift, Kotlin, etc. — are fully supported. Less common languages get partial support (file is indexed but symbol extraction may be limited).

Reasonably-sized files. Files under a few thousand lines are indexed in full. Very large files (typically 5,000+ lines) are summarized rather than embedded line-by-line, both for performance and because embedding quality degrades on huge files.

What's not indexed by default:

Anything in `node_modules`, `vendor`, `dist`, `build`, or other dependency/output directories. These are gitignored by convention and Cursor follows.

Binary files. Images, compiled binaries, fonts, video.

Files explicitly listed in `.cursorignore`. This is the per-project override; Chapter 15 covers it.

Files explicitly excluded by org-wide policy (Enterprise plans only). Admins can configure repository or path blocklists that Cursor will not index regardless of what users want.

What Hits Cursor's Servers

Indexing is partially client-side and partially server-side. Understanding what crosses the wire matters for privacy.

File contents go to Cursor's servers for embedding generation. The server receives the file's text, generates the embedding vector, and stores both the vector and a hash of the content. Plaintext is in transit during embedding; whether plaintext is retained depends on your Privacy mode setting.

With Privacy mode off (default): plaintext may be retained per Cursor's standard data-handling policy. The vectors are retained as part of the index. Cursor's retention policy is documented at cursor.com/privacy.

With Privacy mode on: plaintext is processed in transit and not retained beyond the request. The vector index is still stored (because retrieval requires it), but plaintext is not. This is the right setting for any work on proprietary or regulated code.

Path obfuscation. Cursor obfuscates file paths in some pipelines so the path tree itself doesn't leak structural information about your project. Partial protection — file contents are still visible to embedding generation — but it reduces the surface area of what logs can reveal.

For Enterprise users with stronger requirements: Cursor offers self-hosted indexing options where the index stays on your infrastructure. This is a contractual feature, not part of the standard Pro/Teams offerings.

When the Index Is Slow

Indexing performance scales with codebase size, and large monorepos hit the slow regime. Common situations:

A monorepo with a million files. Initial indexing can take 15+ minutes. Use `.cursorignore` aggressively (Chapter 15) to exclude generated code, build outputs, vendored dependencies, and anything else not actively edited. Excluding 80% of the file count usually drops indexing time by 80%.

A repo with a few very large files. A single 10,000-line file slows indexing more than ten 1,000-line files. If your project has files this large, consider splitting them — but if they need to stay large, the index summarizes rather than fully embeds them, trading index quality for indexing time.

A repo on a slow filesystem. Network filesystems (NFS, SMB), encrypted filesystems with poor I/O, and filesystems with antivirus inspection on every read all slow indexing substantially. The fix is environmental, not Cursor-specific — local SSD with antivirus exclusions for the project directory makes a substantial difference.

A repo that triggers re-indexing on branch switches. Switching branches in git often touches thousands of files. Cursor detects the change and incrementally re-indexes. On a large monorepo, this can mean a noticeable delay after every branch switch. The fix is partly tolerating it, partly using `.cursorignore` to keep the working set smaller.

Re-indexing Triggers

Cursor re-indexes incrementally when files change. The triggers:

File save. Saving triggers a re-index of just that file. Sub-second on most projects.

External file changes detected. When git operations (pull, merge, rebase, checkout) change files, Cursor re-indexes the affected ones. Detection is via filesystem watcher, so it's fast.

Large branch switches. A branch switch that touches many files re-indexes them all. Total time depends on the count.

Manual re-index. Settings → Codebase → Re-index Project forces a full re-index. Useful when something has gone wrong with the incremental index — rarely needed but available.

Incremental indexing rarely needs management. The Merkle tree means unchanged files don't get re-processed, so even big repos stay fast on incremental updates.

Troubleshooting a Slow First-Index

Walkthrough for the most common slow-indexing case: you opened a monorepo, indexing has been running for ten minutes, and Tab feels weak in the meantime.

Step 1: Check `.cursorignore`. Open the file (or create it at the project root if it doesn't exist). Make sure your build outputs, dependency directories, and generated code are excluded. A starter for a typical TypeScript monorepo:

```
node_modules/
dist/
build/
.next/
coverage/
*.lock
*.min.js
*.map
```

Save the file. Cursor picks up the new ignore rules and skips the now-excluded paths.

Step 2: Check the index status. Settings → Codebase → Indexing. The panel shows what's been indexed, what's pending, and any errors. Errors usually point at specific files or directories Cursor can't process.

Step 3: If indexing is still slow after better ignores, force a fresh re-index. Sometimes a partial-but-stuck index keeps Cursor in a confused state. Settings → Codebase → Re-index Project clears it and starts fresh.

Step 4: If it's still slow, look at the largest files in the project. A 50,000-line vendored file in source control is hostile to indexing. Move it out of the codebase, ignore it, or accept the indexing cost.

Step 5: For genuinely huge codebases, consider scoping. If the monorepo is 50 packages and you work on three, configure Cursor to scope indexing to specific subdirectories. The setting is in Codebase → Indexing → Scoped Paths.

Quality Indicators

How can you tell whether your index is good?

Tab proposes codebase-aware completions. When you start writing a function call, Tab fills in arguments that match how the function is used elsewhere. If completions are generic (placeholder names, not your project's actual symbols), the index is incomplete or the file isn't indexed.

Ask answers project-specific questions concretely. "How does authentication work?" should produce an answer with specific file references. If the answer is generic ("authentication is typically handled by..."), the index isn't being consulted properly.

Manual produces coordinated multi-file diffs. When you ask Manual to refactor across files, the proposal references the right files without you attaching them as @ references. If you have to attach files manually for the model to know they exist, the index isn't carrying its weight.

Symbol search works. "Go to Symbol" (Cmd+T/Ctrl+T) should find symbols across the project. This is partially Cursor's indexing and partially VS Code's, but slow or incomplete results suggest indexing trouble.

If your index is bad, the likely causes: (1) `.cursorignore` excluding too much, (2) indexing not yet complete, (3) Privacy mode interacting unexpectedly with cached vectors after a setting change, or (4) very large files exceeding the embed-in-full threshold.

The @ References as a Trust Boundary

The codebase index is the default context source. The @ references in chat (covered in Chapter 6) are an explicit overlay — "consider these specific files in addition to whatever the index retrieves."

The trust hierarchy:

1. **Explicit @ references.** Highest signal. The model sees these in full and treats them as authoritative.
2. **Selection (in Cmd+K, Manual).** The selection is the editable scope; the model treats it as the working surface.
3. **Codebase index retrieval.** The model retrieves what looks relevant. Quality depends on the index.
4. **General training knowledge.** Whatever the model knows from training, when nothing else is loaded.

When the model produces bad output, the fix is usually moving up the hierarchy. If retrieval is missing files you know are relevant, attach them explicitly with @. If selections aren't capturing the right scope, expand or tighten them. The model can only act on what it sees.

Indexing and the Cost Question

Indexing is mostly free at the user level. There's no per-file or per-vector charge in any Cursor plan; the cost of running the embedding pipeline is amortized into your subscription. Re-indexing on every save doesn't drain your credit pool.

What drains credits is what happens after retrieval — when the retrieved context is fed to a frontier model in a chat or Cmd+K request. The frontier model bills per token, and the retrieved context tokens count. A large @ reference attaching a 1,000-line file costs more in chat credits than a focused selection.

The implication: keep your context tight in chat. The index has plenty in it; you don't need to attach everything. Ask, Manual, and Agent will pull what they need from retrieval if you let them. Manual @ attachment is for cases where retrieval misses something specific.

Symbol Search and Navigation

Beyond the AI features, the index also powers Cursor's regular code navigation:

Go to Symbol (Cmd+T/Ctrl+T) — fuzzy-search for any symbol in the project. Powered by the symbol metadata extracted during indexing.

Find All References — given a symbol, find every place it's used. The index makes this much faster than the equivalent search in vanilla VS Code on large projects.

Go to Definition (F12) — navigate from a symbol use to its definition. Works without the AI layer; backed by the language server (TypeScript, Python, etc.) plus the index.

These features make Cursor noticeably faster than vanilla VS Code on large codebases, even before you invoke any AI features. If you're evaluating Cursor and the AI features are a wash for your work, indexed-navigation speed alone may be worth the switch on monorepos.

Chapter 14: Project Rules — `.cursor/rules/*.mdc`

If you only invest in one Cursor feature beyond the basics, make it project rules. Rules are the difference between "the AI sometimes fits my project" and "the AI consistently fits my project." Internal data Cursor has shared from team customers shows acceptance rates climbing from roughly 30% to 80%-plus when teams adopt good rule files — a leverage shift large enough to change what the editor feels like to use.

This chapter is about writing rules that actually shift model behavior, not rules that look good on disk and get ignored in practice. The difference between the two is large and worth understanding before you write a single rule.

What Project Rules Are

Project rules are markdown files in `.cursor/rules/` at your project root. Each file describes a behavior the AI should adopt when working in your project — coding conventions, architectural constraints, naming patterns, security requirements, anything you'd otherwise repeat in every prompt.

Cursor reads the rules into the model's context for every AI invocation in the project. Tab, Cmd+K, chat in any mode, Background Agent, BugBot — all of them see the rules. The rules become a default system-level context that shapes everything the model does in your project, without you referencing them per-prompt.

The file extension is `.mdc` (markdown with frontmatter) by convention, though Cursor reads `.md` files in the directory too. The frontmatter at the top of the file defines metadata — most importantly, which files the rule applies to, via glob patterns.

`.cursorrules` (the legacy format)

If you have an existing project with a `.cursorrules` file at the root, that's the predecessor format — a single text file that applied globally. Cursor still reads `.cursorrules` for backward compatibility, but `.cursor/rules/` is the supported path going forward. Migration is straightforward: split your `.cursorrules` content into focused per-area rule files, place them in `.cursor/rules/`, add appropriate glob scoping to each, and delete `.cursorrules`.

The advantages of `.cursor/rules/` over `.cursorrules`:

- **Per-area scoping.** A rule for your frontend code doesn't apply to your backend code unless you want it to.
- **Per-rule iteration.** Edit, version, and review individual rules without touching the whole rule set.
- **Team review.** Smaller focused files are easier to PR-review than one monolithic file.
- **Rule references.** Rules can reference other files for templates and starter code.

If you're starting fresh, use `.cursor/rules/`. If you're migrating, plan a gradual move rather than a single big-bang change — test that each rule produces the expected behavior before retiring the legacy file.

The Anatomy of a Rule File

A complete rule file looks something like this:

```
---
description: TypeScript style guide for the api/ package
globs:
  - "api/**/*.ts"
  - "api/**/*.tsx"
alwaysApply: true
---

# TypeScript Style - api/

## Imports

- Use named imports, not default imports, for project code.
- Order imports: external libraries, then internal packages, then relative paths.
- No unused imports.

## Types
```

```
- Prefer `type` aliases over `interface` for object shapes.
- All exported functions must have explicit return type annotations.
- Avoid `any` — use `unknown` and narrow with type guards.

## Error handling

- Use the Result<T, E> type from @/lib/result instead of throwing.
- Propagate errors up to the route handler; don't swallow them.
- Route handlers map errors to HTTP responses using the mapError helper.

## Testing

- Each module has a colocated `.test.ts` file.
- Use Vitest, not Jest. Test names follow "should [behavior] when [condition]".
- Mock external services with @/test/mocks; never call real APIs in unit tests.
```

The structure has a few important pieces:

Frontmatter. The block at the top between `---` markers. The fields:

- `description` — a one-line description of what the rule covers. Cursor uses this in the rule picker UI.
- `globs` — file glob patterns that determine when the rule applies. Files matching any glob get the rule in their context.
- `alwaysApply` — if true, the rule applies project-wide regardless of glob (use sparingly; usually you want scoping).

Markdown body. The rule content. Cursor doesn't enforce a specific structure — it's natural-language guidance to the model. Sections, bullets, code examples, references to other files — whatever helps the model understand the behavior you want.

File references. You can reference other project files within a rule (`@template/example.ts`), and those files load into the model's context whenever the rule is active. Useful for "follow this template" rules that point at canonical examples.

What Makes a Rule Good

The temptation when writing rules is to dump everything you wish the AI knew into one giant rule file. That produces rules that look thorough on disk and don't actually shift behavior. The model gets a wall of text that's hard to act on, and the specific guidance gets diluted by surrounding noise.

Good rules have four properties:

Scoped. They apply to a specific area of the codebase, not the whole project. Frontend rules don't apply to backend code; database migration rules don't apply to React components. Use `globs` aggressively. Smaller scope, more focused guidance.

Specific. They give concrete instructions, not vague aspirations. "Use named exports" beats "follow modern conventions." "Mock external services with @/test/mocks" beats "test properly." Specific instructions translate to specific model behavior; vague guidance translates to vague output.

Verifiable. The behavior the rule asks for can be checked, ideally by an automated tool. "Functions over 50 lines should be split" is verifiable (a linter can flag them). "Code should be readable" is not (readability is in the eye of the reviewer). Verifiable rules are also rules the model can self-check before proposing.

Reasoned. They explain *why* the rule exists, briefly. Not a paragraph — a sentence. "Use Result<T, E> instead of throwing because we want errors typed and explicit." The reason helps the model apply the rule to edge cases it didn't anticipate, and helps human readers update the rule when context changes.

Example: A Working Rule

A real rule from a working production project:

```
---
description: Database access patterns for the api/ package
globs:
  - "api/db/**/*.ts"
alwaysApply: false
---

# Database Access — api/db/

## Connection management

- Use the connection pool from `@/db/pool`. Don't create new connections.
```

```
- All queries inside a request handler use the request-scoped transaction
  (available as `ctx.tx`); queries outside a request use `pool.query`.

## Query construction

- Use the Drizzle query builder for new queries. Don't write raw SQL strings.
- Existing raw SQL stays as-is; don't rewrite during unrelated changes.
- Parameterize all user inputs. The query builder does this by default.

## Migrations

- Migrations are append-only. Never modify a migration that's been merged.
- New migrations live in `db/migrations/` named with the timestamp prefix.
- Migrations must be reversible — every `up` has a matching `down`.

## Error handling

- Database errors propagate as `DatabaseError` from `@/lib/errors`.
- Don't catch and log; let the route-level error handler format responses.

## Examples

See `@db/queries/users.ts` for the canonical query pattern.
See `@db/migrations/2026 05 orders.sql` for the canonical migration shape.
```

The rule is roughly 300 words. Narrowly scoped (only files in `api/db/`). Specific instructions (use Drizzle, use `pool` for non-request queries, name migrations with timestamps). Verifiable (a linter can check for raw SQL, the migration naming pattern is regex-checkable). And it points at examples the model can read for ground truth.

When this rule is loaded, the model writes database code that fits the project. Without it, the model writes generally-correct database code that diverges from the project's conventions in ways that show up as friction in code review.

The Glob Scoping Pattern

The biggest predictor of rule effectiveness is the glob pattern. Project-wide rules have to stay generic to apply everywhere; narrowly-scoped rules can be specific.

A practical approach: structure your rules around the actual subdirectories of your codebase. A typical layout:

```
.cursor/
  rules/
    general.mdc                  # Project-wide: language, license, style basics
    frontend.mdc                 # globs: ["app/**/*.tsx", "app/**/*.ts"]
    backend.mdc                  # globs: ["api/**/*.ts"]
    database.mdc                 # globs: ["api/db/**/*.ts"]
    tests.mdc                    # globs: ["**/*.test.ts", "**/*.spec.ts"]
    migrations.mdc               # globs: ["db/migrations/**/*.sql"]
    infra.mdc                    # globs: ["infra/**/*.tf", "infra/**/*.yaml"]
    security.mdc                 # globs: ["**/auth/**", "**/payment/**"]
```

Each file gets specific. The general.mdc covers things that apply everywhere — language choice, the project's license, the fact that you don't use console.log. The per-area files cover area-specific conventions.

The result: when the model edits a file, only the matching rules load. The tests rule fires for tests but not for migrations. The security rule fires for auth and payment code but not for general backend code. The model's context isn't cluttered with irrelevant guidance.

Verifiable Success Conditions

The most powerful rules are the ones tied to verification. The pattern: state what the behavior should be, then state how to check it. The model uses both — the description as guidance and the check as a self-validation step.

Examples:

```
## TypeScript types

- All exported functions have explicit return type annotations.
- Verification: run `npm run typecheck`. New errors block the change.

## Tests

- Every changed function has a test in the colocated `.test.ts` file.
- Verification: run `npm run test:changed` to confirm coverage on changed files.

## Linting

- Code passes the project's lint rules without warnings.
- Verification: run `npm run lint`. Warnings block the change.
```

In Agent mode, these verification steps become part of the agent's loop. After making a change, Agent runs the verification command. If it fails, Agent fixes the issue. If it passes, Agent moves on. Rule-driven verification is what closes the loop on quality.

In Manual mode, verification doesn't run automatically — the model produces a diff, and you run the verification step yourself before merging. The rule still shapes what the model proposes; it just doesn't self-check.

The "30% to 80% Acceptance Rate" Jump

When teams adopt project rules well, the most commonly-cited metric is the change in acceptance rate — the percentage of model proposals (Tab, Cmd+K, Manual) accepted as-is without rejection or revision.

Without rules, baseline acceptance rates in working projects sit around 30–40%. The model proposes plausible code, but it doesn't know your conventions, so a substantial fraction of proposals need editing or rejection.

With well-written rules, acceptance rates climb to 70–80% or higher. Proposals fit the project. Convention drift shrinks. The friction of "rewrite the model's output to match our style" largely disappears.

The leverage compounds because of the volume. If you accept hundreds of Tab proposals a day and your acceptance rate goes from 35% to 75%, you're getting roughly twice as much value per Tab interaction — and ten times the value over a year of project work.

This is why the chapter opened with "if you only invest in one Cursor feature beyond the basics, make it project rules." The investment is real (writing good rules takes hours per area), but the return is large.

A Library of Starter Rules

Appendix B in this book is a 40-plus entry starter library of `.cursor/rules/` files for common stacks — TypeScript/React, Python/FastAPI, Go/gRPC, Rust/Axum, mobile, infra, test discipline, security. The starter library isn't a substitute for project-specific rules, but it's a starting point that captures the conventions most teams in those stacks share.

Use the starters as a base. Drop them into `.cursor/rules/`, run the project for a few days, and refine based on what you see. The first revision pass usually tightens glob scopes and replaces generic conventions with your project's actual ones. After a week or two of iteration, the rules feel like they were written for your project specifically — because they were.

Writing Your First Rule

If you've never written a project rule, the lowest-friction starting point:

Step 1: Pick one area of friction. Where has the model been getting it wrong? Maybe it keeps using `console.log` and your project uses a custom logger. Maybe it keeps creating components in a style that doesn't match the rest of the app.

Step 2: Write a small rule covering that one thing. Three to five sentences. Specific. Scoped to the directory where the friction shows up.

Step 3: Save it as `.cursor/rules/your-rule.mdc`. Cursor picks it up immediately.

Step 4: Use the editor for a day. Notice whether the friction dropped. If yes, the rule works. If no, the rule needs more specificity or wider scope.

Step 5: Iterate. Add a second rule covering a second area of friction. Keep going.

Most teams' rule libraries grow organically over months. Don't try to write a complete rule library in one sitting. Cover what hurts today, then expand as you notice more areas where rules would help.

Rules Hygiene

Patterns to avoid as your rule library grows:

Don't write contradictory rules. If two rules give the model conflicting guidance, the model may follow either, may compromise between them, may follow neither. Read your rules for consistency periodically.

Don't write rules that your codebase doesn't follow. If your rule says "always use named exports" but half the codebase uses default exports, the model gets mixed signals. Either fix the codebase or relax the rule.

Don't accumulate rules without reviewing. A rule that made sense six months ago may be obsolete now. Quarterly cleanup of the rules directory catches rules that no longer apply.

Don't make rules the model can't follow. "All code must be perfectly correct" is not a rule — it's a wish. Rules need to be actionable: "Functions handle the null case," "All API responses include a `correlationId` field." Actionable beats aspirational.

Don't put secrets, credentials, or sensitive data in rules. The rules go to the model on every invocation. Sensitive data does not belong in rule files.

Tighten the Index Next

You have rules. The next concern — especially on large projects — is what's *not* in the index. Performance tuning of indexing on large codebases is its own discipline. Chapter 15 covers `.cursorignore` and the broader performance-tuning story. After that, Chapter 16 covers Background Agents — the cloud-hosted variant of Agent mode that lets you assign work and walk away.

Chapter 15: `.cursorignore` and Performance Tuning

Cursor's defaults are tuned for medium-sized projects — a few thousand files, a few hundred thousand lines of code. They work fine in that regime. Outside it, especially in monorepos with hundreds of thousands or millions of files, the defaults are the difference between an editor that feels fast and one that feels stuck.

This chapter is about the tuning that makes Cursor scale. Most of it is `.cursorignore` — the file that determines what doesn't get indexed. The rest is broader performance hygiene that matters once your codebase is past the easy-defaults size.

What `.cursorignore` Does

`.cursorignore` lives at your project root. The syntax is `.gitignore`-compatible: glob patterns, one per line, with `!` for negations. Files matching any pattern are excluded from Cursor's index. They stay visible to the editor (you can open them, edit them, commit them) but they don't show up in semantic retrieval, don't influence Tab completions, and don't appear in @ references.

`.cursorignore` complements `.gitignore` rather than replacing it. Cursor honors `.gitignore` by default, so anything ignored by git is also ignored by Cursor without an entry in `.cursorignore`. The `.cursorignore` file is for things you want in version control but don't want indexed — generated code, vendored dependencies kept in the repo, large data files.

```
# Build outputs that aren't gitignored
dist/
build/
.next/

# Generated code in the repo
**/generated/
**/*.pb.go
**/openapi-generated/
```

```
# Vendored dependencies
vendor/
third_party/

# Large data files
fixtures/large/
data/snapshots/

# Generated lock files (still relevant to git, not to indexing)
package-lock.json
yarn.lock
pnpm-lock.yaml
Cargo.lock
```

The patterns above cover most of what teams add to `.cursorignore`. The exact list depends on your project's structure and what's in version control.

The Monorepo Problem

Cursor on a monorepo with a million files is a different product from Cursor on a small project. Indexing takes longer, retrieval is slower, ambient memory and CPU load are higher, and completion quality in the parts of the codebase you don't actively edit is weaker.

The fix is almost always `.cursorignore` — aggressive exclusion of files you don't actually edit. Some patterns:

Exclude generated code. Anything in `gen/`, `generated/`, `_pb.*`, `*.gen.*`, generated GraphQL clients, generated Protobuf code, generated OpenAPI clients — none of it benefits from indexing. The generators rebuild it on demand; the index doesn't help you understand it because you wouldn't read it manually anyway.

Exclude tests for parts of the monorepo you don't edit. If the monorepo has 50 packages and you only work on three, the other 47 packages' test files are noise in your index. A glob like `!packages/{my-package,my-other-package}/**/*.test.ts` to exclude all test files except for your packages is heavy-handed but effective.

Exclude old code paths. If the monorepo has a `legacy/` directory representing a deprecated path still in source control for archaeology, ignore it. The model shouldn't propose patterns from legacy code.

Exclude documentation snapshots and build artifacts. Static site builds, generated docs, exported diagrams. These are version-controlled outputs, not editable inputs.

Exclude binaries and assets. Images, fonts, audio, video, compiled binaries. None should be indexed; most are excluded by default but check.

The goal: bring the indexable file count down to the working set you actually care about. A monorepo with a million total files might have an active working set of fifty thousand — the ones you and your immediate teammates edit and read. Indexing fifty thousand files is fast; indexing a million is slow.

Indexing Time Targets

A rough sense of what's reasonable:

Small project (under 5,000 indexed files): Initial index in under a minute. Incremental updates are sub-second.

Medium project (5,000–50,000 indexed files): Initial index in 1–5 minutes. Incremental updates are a few seconds.

Large project (50,000–200,000 indexed files): Initial index in 5–15 minutes. Incremental updates can take 5–30 seconds on big branch switches.

Very large project (200,000+ indexed files): Initial index 15+ minutes. Incremental updates can be slow on big changes. You're in the regime where aggressive `.cursorignore` is necessary, not optional.

If your project's indexing time runs materially longer than these ranges, you have either too many indexed files (use `.cursorignore`), too few resources on your machine (slow disk, low memory), or specific indexing-hostile files (very large files, deeply nested structures with many tiny files). Each problem has a different fix.

The 5,000-Line File Problem

One case worth calling out: very large individual files. A 5,000-line file sits at the edge where Cursor's embedding pipeline shifts from "embed in full" to "summarize." A 50,000-line file is well past it.

The implications of large-file summarization:

Tab autocomplete in the file is weaker. The model has the summary, not the full content. Completions don't reflect the local patterns of the rest of the file as accurately.

Cmd+K rewrites of a selection are weaker. The model needs surrounding context to make a good rewrite; the summary loses it.

Manual edits across files lose precision. If a large file is in scope, the model's grasp of that file is shallower than its grasp of normal-sized files.

Symbol search is unaffected. The file is still indexed for symbols and structure; only the embedding-level semantic understanding is reduced.

If you have files this large, the options:

Split them. A 5,000-line file probably wants to be a directory of 200-line files. The split improves both Cursor's understanding and human review-ability. Often a 5,000-line file signals deferred refactor debt; the move to Cursor is a forcing function to address it.

Accept the reduced understanding. If splitting isn't feasible (generated code, framework-imposed file structure, etc.), Cursor's behavior on that file will be weaker. Chat works; Tab and Cmd+K will be less precise.

Exclude from the index. If a large file is generated and not actually edited, ignore it. You lose Cursor's awareness of the file's contents but drop the indexing cost.

What to Ignore vs. What to Keep

A practical rule: ignore what you wouldn't read.

If you wouldn't open a file to read it manually — generated code, vendored libraries, build outputs, large fixtures — there's no benefit to indexing it. The model proposing patterns from it is more likely to mislead than help.

If you would read a file — your team's source code, the test files for your packages, the configuration that's part of your active work, the docs you maintain — keep it indexed.

The middle ground: dependencies whose source you occasionally read for debugging or comprehension. Most users keep these out of the index (the maintenance friction of indexing third-party code outweighs the rare benefit) and lean on chat with @Web or @Docs references the few times the dependency's behavior matters.

Editor Performance vs. Indexing Performance

Indexing is one performance dimension. The editor's general responsiveness — file open speed, search speed, syntax highlighting, autocomplete latency — is another. Most editor-perf issues in Cursor have the same causes as in vanilla VS Code:

Too many extensions. Each extension runs in its own process and consumes memory; some (especially older ones) are CPU-heavy. Disable extensions you don't actively use.

Too many open editors. Open tabs persist state; on a large monorepo, a hundred open files is a memory drag. Close what you're not actively using.

File watcher saturation. On certain platforms (Linux, especially with low default `inotify` limits), Cursor's file watching can hit OS-level limits on large projects. Raise `fs.inotify.max_user_watches` to 524288 or higher on Linux.

Antivirus inspection. Real-time scanning of every file Cursor opens or watches slows everything noticeably. Add the project directory and Cursor's data directory to your AV exclusions.

Slow filesystem. Network filesystems, encrypted filesystems with poor I/O, and filesystems with many small files all hurt. Local SSD with a normal filesystem is the right baseline for serious development work.

Reducing AI Latency

Beyond indexing and editor responsiveness, the latency you'll feel most is the gap between asking for AI output and receiving it. Some optimizations:

Choose faster models for low-stakes tasks. Tab uses Cursor's tuned fast model and is essentially instant. Cmd+K and chat can run frontier models with longer latency. For tasks where the model choice doesn't materially affect quality (a docstring, a simple refactor), pick a faster model — DeepSeek V4 Flash, GPT-5.4 Mini, or whatever the current "fast tier" options are. Chapter 18 covers the picker in detail.

Keep prompts short. Long prompts with extensive context take longer to process. Concise prompts with focused context run faster and often produce better output.

Tighten selections. A 1,000-line selection has more tokens than a 100-line selection. When the relevant scope is a function, select the function — not the whole file.

Use the right mode. Agent's loop runs multiple turns and takes longer than Manual's single turn. For tasks Manual can handle, the latency penalty of Agent is real.

Cut @ references where the index would do. If the model can find context through retrieval, attaching it explicitly with @ adds no value but adds tokens and latency.

Most users don't need to optimize for latency until they're heavy users; the defaults are fine for daily work. But once you're invoking AI features dozens of times an hour, small savings compound.

Privacy Mode and Performance

A note that comes up often: Privacy mode does not measurably hurt performance. The latency of an embedding request with Privacy mode on matches the latency with it off. The retention policy differs; the path through the infrastructure is largely the same. If you're considering Privacy mode but worried about a perf cost, don't be — the cost isn't there.

What does affect performance under Privacy mode is the absence of cached context across invocations. With Privacy mode off, some optimizations reuse retrieved context across nearby requests; with Privacy mode on, the retention those optimizations require isn't allowed, so each request starts clean. The user-visible difference is usually a small latency hit on rapid-fire requests, not a quality difference. Most users don't notice.

Tuning the Working Set

Some users narrow the working set further — excluding even active code outside the area they're currently focused on. On a monorepo with 50 packages where you're working on one this sprint, you might add exclusions like:

```
packages/!(active-package)/**/*.test.ts
packages/!(active-package)/**/*.stories.tsx
```

This keeps other packages' implementation files indexed (so cross-package references still work) but drops their tests and stories from semantic retrieval. The index shrinks, retrieval speeds up, and your active package's signal isn't drowned by noise from the rest.

The downside: when you switch packages, you have to update the exclusions. For users who rotate quickly, that's too much overhead. For users who stay in one area for weeks, it's a real win.

Tooling Around `.cursorignore`

A few utilities that come up:

Templates. The starter `.cursorignore` files in Appendix F (and at github.com/cursor-community/cursorignore-templates) cover common stacks. Use one as a base and customize.

Per-repo and per-team standards. Larger teams sometimes maintain a shared `.cursorignore` checked in alongside the codebase, so every team member has the same exclusions. This is the right default — `.cursorignore` is project configuration, not user state.

Diagnostic mode. Settings → Codebase → Indexing has a "Diagnostics" view that lists what's indexed, what's not, and why. Useful when you're trying to figure out whether a specific file made it in.

Exclude on demand. Right-click a file or folder in the explorer and select "Add to .cursorignore." Useful for one-off exclusions without editing the file by hand.

A fast index, tuned rules, tight exclusions: the editor scales. The next two chapters extend Cursor's reach beyond the local editor — Background Agents (Chapter 16), where the agent runs on Cursor's infrastructure while you walk away, and `/multitask` (Chapter 17), the Cursor 3.0 feature that runs parallel async subagents in isolated worktrees.

Chapter 16: Background Agents

Agent mode in the foreground is a powerful but supervised tool — you watch it work, intervene when it goes off the rails, and review its diffs. Background Agent is the same primitive moved to the cloud. You kick off a task, close your laptop, go to lunch, and return to a finished branch and a draft PR. The agent worked while you were away.

This is the feature that pushed Cursor from "smart editor" into "agent runtime" territory. It's also the feature most prone to misuse — Background Agents working autonomously on tasks they shouldn't have been given produce expensive messes. This chapter is about getting the most out of Background Agents while staying out of the failure modes.

What Background Agents Are

A Background Agent is an Agent run that executes on Cursor's cloud infrastructure rather than in your local editor. The infrastructure provides:

- A sandboxed environment with your project's code (cloned from the repo or pushed up from your local checkout)
- A shell environment configured for your project (Node, Python, Go, whatever your project needs, configured per `.cursor/environment.json`)
- The same model picker, project rules, and MCP integrations as the local editor
- A persistence layer that records the agent's actions and a UI to review them later

When you kick off a Background Agent, you describe the task in a prompt — usually a ticket-shaped prompt much like what you'd send to a foreground Agent. The cloud infrastructure spins up a sandbox, clones the repo into a worktree, and starts the agent loop. The agent works for anywhere from a few minutes to an hour, depending on the task. When it finishes (or hits a stopping condition), it pushes a branch and opens a draft PR for review.

The whole interaction is asynchronous. You don't have to keep your editor open. You don't have to watch. You get a notification when the run completes, click through to review, and decide whether to merge, request changes, or discard.

When Background Agents Are the Right Tool

Background Agents shine when the task fits four conditions:

The task is well-scoped enough to run unsupervised. You can describe the work clearly, with enough specificity that an agent could execute it without follow-up clarification. "Add a `created_at` timestamp column to the orders table and plumb it through the API and frontend" is well-scoped. "Improve the orders code" is not.

The task is large enough to justify the setup cost. Spinning up a Background Agent has overhead — sandbox cold-start, clone, environment setup. For a five-minute task, you'd usually just do it locally. For a thirty-minute task, the parallelism is worth the overhead.

The task can be verified after the fact. The agent's output is a PR with tests, a description, and a diff. You should be able to verify the work from the PR alone, without debugging the agent's reasoning. Tasks where verification means reading thirty step-by-step decisions are poor Background Agent fits.

The cost of a wrong result is bounded. Background Agents work on a branch, not on main. Worst case, you discard the branch and spend some credits. If the cost of a wrong direction is "we don't merge the PR," Background Agents are fine. If the cost is "the agent broke something subtle that ships before we notice," the task wants foreground supervision.

Concrete examples of good Background Agent tasks:

- **Ticket implementation.** "Implement issue #1234 according to its description. Run all tests. Open a PR."
- **Test coverage backfilling.** "Add unit tests to the `payments/` package to bring coverage from 60% to 85%. Tests should be meaningful, not assertion-padding."

- **Dependency upgrades with code changes.** "Upgrade Express from 4.x to 5.x in the api package. Apply any breaking-change migrations. Run tests."
- **Mechanical refactors.** "Convert all class components in the `app/components/` directory to function components."
- **Documentation generation.** "Add JSDoc to every exported function in the `lib/` directory. Match the existing JSDoc style."

These tasks share a shape: concrete, explicit success criteria, reviewable as a diff.

When Background Agents Are the Wrong Tool

Conversely, Background Agents are the wrong tool when:

The scope is fuzzy. "Make the search faster" or "clean up the auth code" don't have a clear target. The agent will pick an interpretation and execute it; you'll likely get something that doesn't match what you wanted.

The task requires mid-execution decisions. "Refactor the database layer" raises questions the agent will answer for you (which patterns to standardize on, how to handle the migration, what to do with edge cases) – and you may disagree with its answers. Tasks with mid-execution decisions belong in foreground Agent or Manual, where you can intervene.

The task is exploratory. "See if it's feasible to switch from REST to gRPC." That's investigation, not implementation. Use Ask in the foreground.

The result needs continuous validation. Tasks where you'd want to step through changes one at a time, verifying as you go, don't fit the asynchronous model.

The credit cost is unbounded. A task that could plausibly take an agent two hours of premium-model time, with no certainty of success, is an expensive bet. Background Agents bill the same as foreground; long failed runs are real money. Cap the scope or pick a cheaper model.

Kicking Off a Background Agent

The flow:

Step 1. From the chat panel or via the command palette, choose "Run as Background Agent." (The exact UI changes; the capability is consistent.)

Step 2. Compose the prompt. It should be detailed — more so than a foreground Agent prompt, because there's no opportunity for follow-up clarification. Include:

- The goal in clear terms
- The success criteria
- Any constraints (don't change unrelated code, preserve specific behavior, use specific patterns)
- The verification commands the agent should run (tests, lint, type-check)
- The expected output format (a branch and a PR, with a description that...)

Step 3. Choose the model. For Background Agents, model choice matters more than for foreground because no human is checking each step. Premium models (Claude Opus 4.7, GPT-5.4) produce more reliable autonomous output. The cost is real; for serious tasks, it's usually worth it.

Step 4. Choose the branch starting point — usually `main` or a feature branch you're building on. The agent works from a worktree at that branch.

Step 5. Send. The Background Agent run starts in the cloud. You can close the editor; the run continues.

Step 6. Wait for the notification. Cursor pings you (in-app, optionally email or Slack if configured) when the agent finishes or hits a stopping condition. The run also shows up in your Background Agent dashboard at any time.

Step 7. Review. Open the resulting PR. Read the diff. Run the tests yourself if you want a second confirmation. If the work is good, merge. If it needs revision, comment on the PR or kick off a follow-up run with the corrections.

Writing Good Background Agent Prompts

Background Agent prompts are longer and more specific than foreground Agent prompts. Three to five paragraphs is about right. The structure that works:

Paragraph 1: The goal. What are you trying to accomplish? Why? "Implement the new admin reporting endpoint described in issue #4567. The endpoint returns aggregated weekly metrics for the admin dashboard."

Paragraph 2: The scope. What files and behavior are in scope? "The endpoint lives at `GET /api/admin/reporting/weekly` and is implemented in `api/admin/reporting.ts`. It returns the schema described in the issue. The handler should use the existing `MetricsService` for data access."

Paragraph 3: The constraints. What's out of scope? What should be preserved? "Don't change the `MetricsService` interface — only call existing methods. Don't touch any unrelated routes. Use the project's standard error handling. Match the existing pagination conventions."

Paragraph 4: The verification. What should the agent run to confirm the work? "Run `npm run typecheck`, `npm run lint`, and `npm test` in that order. All must pass. Add unit tests for the handler in `api/admin/reporting.test.ts` covering happy path, empty data, and pagination edges."

Paragraph 5: The deliverable. What's the expected output? "When done, push a branch named `agent/admin-weekly-reporting` and open a PR titled 'Add admin weekly reporting endpoint' with a description that summarizes the implementation and links to issue #4567. Include a short note on any decisions you made that weren't covered by the prompt."

A prompt like this gives the agent enough scaffolding to execute autonomously without surprises. The verbosity isn't fluff — it stands in for the back-and-forth you'd have with a foreground Agent.

The PR Review Loop

When a Background Agent's PR shows up for review, treat it like a junior engineer's PR — with the added caveat that the agent doesn't notice things you're not asking about.

The review checklist:

Did the agent do what was asked? Read the description; check that the diff matches. If the description says "added pagination" but you don't see pagination code in the diff, something's off.

Are the tests meaningful? Background Agents sometimes add assertion theater — tests that pass because they're underspecified, not because the code is right. Read them and check that they verify behavior.

Did the agent stay in scope? Look for unrelated changes the agent made along the way. Sometimes the cleanup is good; sometimes it's scope creep that should be reverted.

Are the verification steps green? CI should run the verification commands you specified. If CI fails, the agent's claim of "tests pass" is wrong; investigate before merging.

Did the agent handle edge cases? Pagination with zero results. Empty input. Auth failures. The kinds of cases careful engineers think about and rushed engineers don't. Background Agents land on the rushed-engineers side more often than not; explicit edge-case tests help.

Are the docs updated? If the change should have updated the API docs, the README, or the CHANGELOG, did the agent do it? Often not, unless the prompt said so explicitly.

If the PR is good, merge. If it needs minor changes, comment on the PR (Cursor's GitHub integration lets the agent see comments) or open it locally and finish by hand. If it's substantially wrong, discard the branch and reconsider whether the task was well-scoped enough for a Background Agent.

Parallel Background Agents

You can run multiple Background Agents simultaneously. The dashboard shows all active runs; each operates on its own worktree and produces its own PR. This is where Background Agents become a workflow multiplier — kick off three tickets, do other work for thirty minutes, return to three PRs.

Patterns that work well with parallelism:

Independent tickets. Three different tickets in three different parts of the codebase. The PRs don't conflict; review them independently.

The same task in different places. "Add the new logging convention to the auth package," "Add the new logging convention to the payments package," "Add the new logging convention to the admin package" — three agents doing structurally similar work in non-overlapping codebases.

Independent test coverage tasks. Three packages all needing test coverage backfilling. Each agent works in its package; the PRs don't interfere.

Dependency upgrades by area. When a major version bump touches multiple packages, splitting the work across agents lets each focus on its package's specifics.

What doesn't work well with parallelism:

Tasks that touch the same files. Two agents modifying the same files produce conflicting branches. You'd merge one and rebase the other — not impossible, but adds friction.

Tasks with shared review burden. Five PRs to review at once is real work even if each PR is small. Parallelism that produces a review backlog you can't keep up with is leverage you can't use.

Tasks that depend on each other. "Refactor the database layer" and "use the refactored database layer in the API" are sequential, not parallel. The second can't run usefully until the first is merged.

Background Agent vs. Foreground Agent

When to use which:

Foreground Agent — interactive, you watch and intervene, lower setup cost, better for ambiguous or evolving tasks. Use when:

- You want to make decisions during execution
- The task is unclear enough that you need to watch
- You're working on something high-stakes where step-by-step review matters
- You're using Cursor for the next thirty minutes anyway

Background Agent — asynchronous, runs without supervision, higher setup cost, better for clear tasks. Use when:

- The task is well-scoped and verifiable
- You'd rather work on something else for the next thirty minutes
- The result is a PR that someone (you or a teammate) reviews like any other PR
- Running multiple in parallel is a win

Most users find the right ratio is something like 70% foreground/30% background. Background Agent is leverage when the task fits, but most daily tasks don't fit. Don't push tasks into Background Agent that don't belong there.

Cost Considerations

Background Agents run on premium models for the most part (the autonomous loop benefits from frontier capability). They run for tens of minutes. They consume real credits. A typical Background Agent run on Claude Opus 4.7 costs $2–5 in credits; long runs cost more.

The cost calculus:

- A 30-minute agent run that produces a working PR is worth more than 30 minutes of your time at almost any salary.
- A 30-minute agent run that produces a PR you have to throw away is pure waste.

The skill is keeping the success rate high. Good prompts, well-scoped tasks, and the discipline to recognize when a task isn't fit for unsupervised execution are what separate productive Background Agent users from frustrated ones who burn credits and get nothing.

The `.cursor/environment.json` File

Background Agents need to know how to set up your project's environment — language version, package manager, setup commands.
The `.cursor/environment.json` file at your project root captures this:

```
{
  "setup": [
    "nvm use 20",
    "npm ci",
    "cp .env.example .env"
  ],
  "verify": [
    "npm run typecheck",
    "npm run lint",
    "npm test"
  ],
  "language": "typescript",
```

```
  "packageManager": "npm"
}
```

The `setup` commands run when the sandbox is provisioned. The `verify` commands are what the agent runs to confirm the work is correct. Cursor uses these defaults; agents can run additional commands as the task requires.

For projects without an `environment.json`, Cursor infers the setup from common config files (`package.json`, `pyproject.toml`, `Cargo.toml`, etc.). For complex projects with non-standard setup, the explicit file produces more reliable Background Agent runs.

Chapter 17: `/multitask` and Async Subagents

`/multitask` is the headline feature of Cursor 3.0. The slash command spawns multiple async subagents that work in parallel on the same task, each in an isolated worktree, then surfaces the results so you can pick the best one. The mental model is "try three approaches at once and keep the winner" — turned into a primitive instead of a workflow you orchestrate by hand.

This chapter covers what Multitask does, when it wins over a single Background Agent or foreground Agent run, and the discipline needed to keep it from becoming a chaos generator.

What Multitask Does

When you invoke `/multitask` in chat, you describe a task and specify how many parallel agents should work on it. Cursor spawns N copies of the agent infrastructure, each:

- Runs in its own isolated git worktree (a separate filesystem checkout pointing at a separate branch)
- Has its own model selection (you can specify the same model for all subagents or different models per subagent)
- Works through the task independently, without seeing what the other subagents are doing
- Reports back when finished

The output is a multi-pane review interface: each subagent's branch shown side-by-side with diffs, commit history, and test/verification output. You compare, pick the winner, and pull that subagent's work into your foreground branch. The losing subagents' worktrees are discarded (with their branches optionally retained for archaeology).

The point of the parallelism is exploration. Different subagents may pick different approaches — one refactors incrementally while another rewrites cleanly; one adds a new abstraction while another reuses an existing one; one model's reasoning converges faster on a task than another's. Running multiple lets you compare without committing.

Why Worktrees Matter

Worktrees are git's mechanism for having multiple working copies of the same repository, each on a different branch, sharing the same `.git` directory. Multitask uses worktrees rather than separate clones because they're lightweight (no full re-clone per subagent) and keep the agents isolated from each other's filesystem state.

The implications:

No file-level conflicts during execution. Each subagent has its own working tree. If two subagents edit the same file, they don't trip over each other; their edits land on different branches.

Branch-level conflicts at merge time. When you pick a winner, you're picking a branch. Merging that branch back to your foreground branch follows normal git merge rules. If your foreground has changed during the run, you may have a normal merge conflict to resolve.

Independent test runs. Each subagent runs the verification commands in its own worktree. One subagent's failing tests don't affect another's.

Faster than separate clones. A worktree shares the `.git` directory with the parent checkout, so creation is fast (seconds, not minutes for a large repo).

The end-user experience hides most of this; you don't manage worktrees directly. But understanding that subagents are isolated by worktrees explains why Multitask scales to several parallel agents without them getting tangled.

When Multitask Wins

Multitask is the right tool in three situations:

You're not sure which approach is best. "Refactor the search service to be faster" has multiple valid approaches — caching, query optimization, indexing changes, algorithmic changes. Running three subagents with different prompts ("focus on caching," "focus on query optimization," "focus on indexing") and comparing results beats picking one approach upfront, finding it doesn't work as well as you hoped, and switching.

You want to compare model strengths. Run the same task on Claude Opus 4.7, GPT-5.4, and DeepSeek V4 Pro. Compare the outputs. Different models genuinely produce different code on substantive tasks, and on a task where the model choice matters, Multitask is the cheapest way to compare.

The task has high variance. Some tasks have a luck-of-the-draw element — a single agent run might converge well or wander into a bad direction. Running multiple in parallel and picking the best hedges against the bad-run case. Especially valuable when the cost of a bad direction is large (wasted review time) and the cost of running extra is small (credits).

When Multitask Adds Chaos

Multitask is the wrong tool when:

The task is well-defined enough that all subagents produce similar output. "Rename `getCwd` to `getCurrentWorkingDirectory`" doesn't benefit from parallelism — three subagents do the same rename, and you get three nearly-identical PRs. Running one is just as good and costs a third as much.

You don't have time to review three results carefully. Multitask's win depends on careful comparison. If you're going to glance at three branches and pick the first one that looks plausible, you're not getting the value — you're paying 3x for the same outcome as a single run.

The cost is unbounded and you're not budgeting for it. Running three Opus 4.7 subagents on a complex task costs $6–15 in credits. If each plausibly takes an hour, the math gets uncomfortable. Use Multitask deliberately, with the budget in mind.

The decision is mid-task, not post-task. Multitask compares finished outputs. If the right approach depends on intermediate results — "we should know after the migration runs whether this approach is feasible" — sequential execution with foreground supervision fits better.

The tasks would conflict at merge time. Three subagents each editing the same authentication code produce three conflicting branches. The "pick the winner" step is fine; what's costly is spending three runs of credits on outputs that mostly overlap when only one will be used.

Pulling the Winner Into the Foreground

After Multitask completes, the review interface lets you compare the subagent outputs and pick which to keep. The selected subagent's work pulls into your foreground branch via a normal git merge or rebase (Cursor handles the mechanics). The other subagents' branches are typically discarded — their `.git` references cleaned up — but Cursor lets you retain branches if you want to review them later.

The pull-into-foreground step is where you commit to one direction. After it, you're back in the normal Cursor workflow: review the diff, run tests yourself, refine, merge.

A subtlety: if you've done other work in your foreground branch during the run, the pull may produce a normal merge conflict. The resolution is the same as for any branch merge — Cursor's diff-resolution UI works the same way whether the conflict is Multitask-driven or otherwise.

Walkthrough: Trying Three Approaches in Parallel

Concrete scenario. Task: "The `searchProducts` function is slow for queries over 1000 results. Make it faster."

You're not sure whether the right answer is caching, indexing, query optimization, or pagination redesign. Multitask fits.

Step 1: Compose the multitask prompt.

In chat, type:

> "/multitask 3
>
> The `searchProducts` function in `services/search.ts` is slow for queries returning over 1000 results. Make it faster while preserving the current API contract.

- Subagent 1 (Claude Opus 4.7): Focus on caching. Add a result cache with appropriate invalidation.

- Subagent 2 (Claude Opus 4.7): Focus on query optimization. Look at the underlying database queries and improve them.
- Subagent 3 (Claude Opus 4.7): Focus on pagination and lazy loading. Don't fetch all results upfront.

Each subagent: run `npm test` to confirm tests pass, run `npm run benchmark:search` and include the result in the PR description. Open a PR titled 'Search optimization (approach: ...)' and describe the approach."

Step 2: Send. Three subagent runs start in parallel. Cursor shows the dashboard with three live runs.

Step 3: Wait. This takes 10–30 minutes depending on the complexity of each approach. Do other work; Multitask is asynchronous.

Step 4: Review the dashboard. When all three complete, the dashboard shows:

- Subagent 1: Branch `multitask/search-cache-1`. Tests pass. Benchmark shows 60% improvement on the hot query, 5% on cold queries.
- Subagent 2: Branch `multitask/search-query-1`. Tests pass. Benchmark shows 40% improvement across the board, smaller variance.
- Subagent 3: Branch `multitask/search-pagination-1`. Tests pass with one new test added. Benchmark shows 20% improvement on initial load but markedly faster perceived client experience (lazy loading).

Step 5: Compare. Open each branch's diff. Read the code. Run the benchmark yourself to verify.

Step 6: Pick. Subagent 2's approach is the most balanced — improves all queries without adding cache invalidation complexity. Pull it into your foreground branch.

Step 7: Refine. Maybe you take an idea from subagent 1 (the cache pattern) and layer it on top of subagent 2's query optimization. Or maybe you ship subagent 2's approach as-is. Either way, you're not stuck with a single agent's output.

The whole exercise might cost $15–25 in credits. You get a deeper exploration of the design space than a single Agent run would produce, and you end up with a PR better than any single subagent would have produced alone.

Multitask vs. Background Agent vs. Foreground Agent

A summary of when each fits:

Foreground Agent — supervised, single run, lower cost. Use for ambiguous or evolving tasks where you want to intervene. Most daily work.

Background Agent — autonomous, single run, higher latency. Use for well-scoped tasks that benefit from running while you do something else.

Multitask — parallel, multiple runs, higher cost. Use when the choice of approach is genuinely uncertain and worth exploring.

The decision flow:

1. Is the approach clear? → Foreground or Background Agent.
2. Is the approach unclear and worth exploring? → Multitask.
3. Will I be at the keyboard for the next 30 minutes? → Foreground.
4. Will I be doing something else? → Background.

Most days, Multitask doesn't come up. When it does — usually for an open-ended performance optimization, a refactoring decision, or a task where you want to compare models — it's the right tool for what would otherwise be hours of sequential trial-and-error.

Configuring Subagent Diversity

The default `/multitask N` runs N subagents with the same model and prompt. For exploration tasks, that's wasted parallelism — three subagents running the same prompt with the same model converge to similar outputs.

The diversity comes from varying:

Models per subagent. "Subagent 1: Claude Opus 4.7. Subagent 2: GPT-5.4. Subagent 3: DeepSeek V4 Pro." Different models genuinely produce different code on substantive tasks; comparing across models is one of the best uses of Multitask.

Prompt focus per subagent. As in the walkthrough above — give each subagent a different aspect of the problem to optimize for. The shared goal stays the same, but the angle of attack differs.

Constraints per subagent. “Subagent 1: no new dependencies. Subagent 2: free to add dependencies.” “Subagent 1: minimal change. Subagent 2: ambitious refactor.” Different constraints produce different solutions.

Starting branches per subagent. Most Multitask runs start from the same branch. Occasionally you want different starting points — one subagent on main, another on a feature branch — to compare how the task lands in different contexts.

The skill of Multitask is composing the diversity. Without it, you’re paying for parallel runs that produce similar output. With it, you get genuine exploration.

Multitask Limitations and Failure Modes

A few things that go wrong with Multitask runs:

One or more subagents hang or fail. Sandboxing issues, model timeouts, environmental problems. The dashboard shows failed runs; restart or write off as needed. Failed runs still consume some credits but not the full price of a successful run.

All subagents converge to similar output. Diversity wasn’t enough. Either accept the convergence (one of them is the answer) or rerun with more variation in prompt or model.

The subagents diverge wildly and none is good. The task wasn’t well-scoped enough. Restart with a tighter scope and possibly fewer subagents.

Cost runs higher than expected. Some subagents loop on a hard problem and rack up credits. Multitask has per-run timeouts, but not all runs time out cleanly. Watch the dashboard during long runs and intervene if a subagent is stuck.

Review takes longer than expected. Three full PRs to read is real work. If the comparison takes longer than the runs themselves, the tasks weren't worth running in parallel — sequential exploration would have been cheaper.

The general principle: Multitask amplifies high-quality decision making. It doesn't help if you're not investing in the comparison step.

From Toolkit to Choices

Part IV is done. You have the full toolkit for working at scale: a healthy index, focused rules, tight ignores, Background Agents for solo asynchronous work, Multitask for parallel exploration. Part V shifts to the model layer — the picker, credits and pricing, BYOK, and Privacy mode. Chapter 18 starts with the picker: when each of the four frontier models is the right pick, and how to build the intuition that turns "always use Opus" into "use Opus when it matters."

Part V: Choosing and Using Models

"The expensive model is not the right model for everything, and the right model is not always the expensive one."

— A working principle

Chapter 18: The Model Picker

The model picker is the dropdown most Cursor users glance at once during setup and never touch again. That's a mistake. Choosing the right model for the task is the single biggest lever you have over both quality and cost. Always-Opus is expensive and slow; always-cheap-model runs faster but produces output you'll waste time fixing. The fluent middle — different models for different tasks — separates a Cursor workflow that feels fast and cheap from one that feels slow and expensive.

This chapter covers each of the four frontier models in the picker, what each is best at, and the intuition for picking among them.

What's in the Picker

As of early 2026, Cursor's model picker offers four frontier models, each with one or more variants, plus Cursor's own fast model for Tab. The four frontier options:

Claude Opus 4.7 (Anthropic). The top of the picker for many tasks. Strong on architectural reasoning, gnarly debugging, careful refactors, long-context work. Highest cost per token of the four. Cursor exposes Opus 4.7 with a 1M-context variant; the standard variant has shorter context.

GPT-5.4 (OpenAI). The safe broad default. Strong across most workloads, well-rounded, very fast on shorter tasks. Mid-to-high cost. The GPT family updates more frequently than the others, so the specific variant in the picker shifts; the rough position in the lineup is "the reliable generalist."

DeepSeek V4 Pro (DeepSeek). The cost-quality leader. Roughly comparable to Opus on most coding workloads at roughly one-tenth the cost. The frontier reasoning option for cost-conscious teams. Two variants: V4 Pro (larger, smarter) and V4 Flash (smaller, faster, cheaper). DeepSeek V4 Pro joined the picker in April 2026.

Gemini 3.1 Pro (Google). The long-context champion. 1M+ token context window standard. Strong on tasks where the relevant context spans many files. Mid-cost.

Plus the cheaper tier:

Cursor's own fast model. Powers Tab autocomplete and some default Cmd+K actions. Free at the per-invocation level on paid plans (subject to fair-use). Fast, codebase-aware, tuned for the completion task. Not appropriate for chat or Agent.

The picker also includes "Auto" — Cursor selects per task based on heuristics. Auto is reasonable for users still building intuition; experienced users mostly bypass it for explicit choice.

The Four-Model Mental Model

A rough one-line summary I find useful for daily picking:

- **Opus** when the model needs to think.
- **GPT** when the task is broad and uncertain.
- **DeepSeek** when cost matters and the task is concrete.
- **Gemini** when the context is large.

That's coarse. The next sections refine it.

When to Pick Claude Opus 4.7

Opus is the model to reach for when depth of reasoning matters more than speed. Specifically:

Architectural decisions and design choices. "Should this be a service or a library?" "Where should this validation live in the stack?" "What's the right abstraction here?" Opus's reasoning produces more coherent design discussions than the alternatives — it tracks tradeoffs more cleanly and surfaces constraints the others miss.

Hard refactors that touch many concerns. Refactors that require holding multiple constraints at once — preserve all behavior, don't break this test, follow this convention, watch out for that edge case. Opus juggles constraints better than the others on substantive tasks.

Debugging gnarly issues. When a bug has no clear cause and you need a model that can hypothesize, test, and iterate carefully. Opus produces more useful diagnostic chains than GPT or DeepSeek on cases where the obvious fix isn't right.

Careful code review. When you want a model to read a PR and tell you what's wrong, Opus catches subtler issues than the alternatives. BugBot itself runs on Anthropic models for this reason.

Tasks where wrong is expensive. Authentication, billing, data integrity, security boundaries. Opus's marginal quality is worth the marginal cost when the cost of being wrong is real.

When Opus is the wrong choice: small Tabs, simple Cmd+K rewrites, anything well-defined where the model already has plenty of context. You're paying for reasoning capacity you don't need.

When to Pick GPT-5.4

GPT is the model to reach for when you want a strong generalist that handles most tasks well without specialty consideration. Specifically:

General-purpose chat exploration. "Walk me through how this codebase works" or "What's the right way to test this?" — questions where the task isn't a particular specialty and you want a confident, broad answer. GPT's chat fluency is high.

Tasks heavy on natural-language reasoning. PR descriptions, design docs, explanatory comments, release notes. GPT's prose reads more naturally than the others' on long-form English output.

Quick iteration cycles. GPT is generally fast — typically faster than Opus on equivalent prompts. For tasks where you'll do many rounds of iteration, the latency adds up.

Agent workflows where the task is broad. GPT handles long autonomous loops well; the failure modes differ from Opus's but the success rate on well-scoped Agent tasks is comparable.

When GPT is the wrong choice: tasks where Opus's deeper reasoning matters, tasks where DeepSeek's cost advantage matters, tasks where Gemini's long context matters. GPT is the safe default; the specialty models beat it in their specialties.

When to Pick DeepSeek V4 Pro

DeepSeek is the model to reach for when cost matters and the task is concrete. Specifically:

Volume work where unit cost compounds. Tab is free; chat and Agent aren't. If you're invoking the chat tier dozens of times a day, the difference between Opus pricing and DeepSeek pricing over a month is hundreds of dollars at the team level. DeepSeek lets you afford the volume.

Concrete, well-scoped tasks. "Implement this function," "Add tests for this module," "Refactor this class to use composition" — tasks mechanical enough that the model's depth of reasoning matters less. DeepSeek's quality on these is comparable to Opus or GPT at a fraction of the cost.

Reasoning-heavy tasks where you want frontier capability without frontier price. DeepSeek V4 Pro's thinking mode is genuinely strong on math, complex algorithms, and reasoning chains. For these tasks, you get Opus-comparable output at a fraction of the cost.

Tasks where you'll do many rounds. Iterative debugging, Multitask exploration, repeated agent runs while you tune a prompt. The cost compounds across rounds, and DeepSeek keeps the budget reasonable.

Self-hosted comparison baseline. DeepSeek's open-weight availability means teams running DeepSeek themselves (via Ollama, vLLM, BentoML — covered in *DeepSeek: The Complete Guide*) can use Cursor's BYOK feature to point at their own deployment. Free at inference time, controlled latency, no Cursor credits consumed.

When DeepSeek is the wrong choice: tasks where Opus's additional reasoning capacity genuinely matters and the cost is justified. The DeepSeek-vs-Opus gap is small but real on hard tasks; for those, Opus is still worth the extra cost.

When to Pick Gemini 3.1 Pro

Gemini's edge is context length. The model handles 1M+ tokens of context comfortably, where Opus's standard variant tops out lower. Specifically:

Tasks spanning many files. When the relevant context is a dozen files totaling 50,000+ lines, Gemini handles the load without forcing you to pick which files to attach. The other models would truncate, summarize, or miss context.

Reading large documents. API specifications, RFCs, design docs, PDFs of standards — long single-document tasks. Gemini's context window swallows the whole thing and reasons over it.

Cross-package monorepo work. Tasks that touch a substantial chunk of a monorepo and need awareness across packages. Gemini's context capacity is the difference between "the model sees the relevant context" and "the model sees a small slice and guesses."

Agent tasks with long working state. As an Agent run progresses, working state accumulates — file contents, command output, planning notes. Long agent runs hit context limits in shorter-window models; Gemini's window is generous enough to hold longer runs without summarization-driven quality loss.

When Gemini is the wrong choice: short tasks. The advantage lies in the long-context regime. For quick edits, Gemini doesn't beat the alternatives, and latency on long-context invocations runs higher.

Auto: When to Use the "Just Pick" Setting

Cursor's Auto setting picks a model per-invocation based on heuristics — task type, prompt length, mode, and other signals. Auto is reasonable when:

You're new to Cursor. Building intuition about which model fits which task takes weeks. Auto is good enough during the learning phase to defer the decision.

You're doing routine work. Daily Tab, Cmd+K, and small chat tasks where the model choice doesn't materially affect the outcome. Auto picks reasonably.

You don't want to think about it. For users who'd rather not invest in model-picking discipline, Auto is fine. The cost runs marginally higher than careful manual picking, but the time saved on decisions can be worth it.

When Auto is the wrong choice:

Tasks where the model genuinely matters. Hard debugging, careful refactors, anything where you'd specifically want Opus's depth or Gemini's context. Auto might pick the cheaper option; that's fine for routine work but wrong here.

Cost-conscious workflows. Auto leans toward "premium model when uncertain." If your priority is keeping credits low, manual selection toward DeepSeek is cheaper.

Workflows optimizing for a specific outcome. Multitask runs, A/B comparisons of approaches, anything where you want to control variables. Auto's variability is unhelpful.

The book's recommendation: use Auto for the first month. After that, you'll have intuition about which model fits which task. Switch to manual selection then; you'll save credits and get better quality per dollar.

The Same Task Across All Four

A useful exercise once a month: take a substantive task you'd normally Cursor your way through, and run it across all four models in Multitask. Read the four results. Notice the differences.

The exercise builds calibration in a way nothing else does. You see, concretely:

- Where Opus's reasoning produces a meaningfully better diff than GPT's
- Where DeepSeek matches Opus at a tenth the cost
- Where Gemini's long-context awareness catches something the others miss
- Where the differences are smaller than you assumed

The output isn't a single decision; it's an updated mental model of when each model earns its keep. After three or four such exercises across different task types, your picker discipline sharpens without conscious effort.

Switching Models Per Invocation

Cursor lets you change the model per request via the picker dropdown in the chat panel or via inline prompt syntax. Switching is free (no warm-up, no setup) — use it freely.

Some patterns:

Default to a moderate-cost model, escalate when needed. Set your default to GPT-5.4 or DeepSeek V4 Pro. Escalate to Opus when the task is hard. De-escalate to Cursor's fast model for trivial work.

Match the model to the mode. Tab uses Cursor's fast model. Cmd+K can use any frontier model; pick by task complexity. Manual usually wants a strong model for diff quality. Agent should usually use a frontier model — agent loops degrade with cheaper models.

Differentiate by task class. Refactors → Opus. Generation → GPT or DeepSeek. Long-context exploration → Gemini. Quick reformatting → DeepSeek Flash or fast model.

Switch mid-task if needed. A chat session can change models between turns. Start with DeepSeek for cheap exploration, switch to Opus for precise implementation, switch back to DeepSeek for test generation.

A Daily Picker Routine

The picker discipline that produces consistent quality without unconsciously burning credits:

Morning: review yesterday's credit usage. Notice which model consumed the most and ask whether it was the right choice. Adjust expectations for today.

Per-task: before sending a request, take a half-second on the picker. Default fingers reach for Opus; the half-second decision often picks DeepSeek or GPT for what doesn't actually need Opus.

During Agent or Background Agent runs: Opus or GPT-5.4 unless the task is concrete enough for DeepSeek. Cheaper models in agent loops produce noticeably worse output.

Weekly: check usage. On track for the month? If not, where's the spend going? Adjust.

This routine takes minutes per day and saves substantial credits over a month — which is the bridge to Chapter 19, where picker decisions get translated into dollars: credits and usage-based pricing, the "$20 in three days" trap, and the workflow discipline that prevents it.

Chapter 19: Credits and Usage-Based Pricing

Cursor's pricing model changed in mid-2025 from request quotas to credit pools, and the change has been the most contentious thing the company has done. Understanding how the credit pool works — and how to keep your monthly bill from running away — is the second most important skill in Cursor after the mode rubric. This chapter is about getting the cost story right.

The 2025 Pricing Shift

Pre-2025: Cursor Pro included a fixed number of "fast requests" per month (500 was the long-running number) plus unlimited slow requests. The model was simple: pay $20, get 500 fast requests, the rest are slow. Heavy users sometimes ran out of fast requests and tolerated slow ones for the rest of the month.

Mid-2025: Cursor moved to credit-pool pricing. The plan now includes $20 of usage credits per month at standard per-token rates. Different models cost different amounts; different actions consume different numbers of tokens. Heavy users no longer get unlimited slow requests as a backstop; they hit the credit cap and get either a hard stop or pay-as-you-go billing.

The change was contentious for several reasons:

Costs got less predictable. Fixed-quota pricing was easy to understand. Credit-pool pricing means a single Opus 4.7 Agent session can consume more credits than a hundred small chat queries.

Heavy users felt it disproportionately. Some went from "occasionally hitting the slow tier" to "running out of credits on day three." The new model penalizes heavy use much more than the old one did.

The math was opaque at first. Per-token rates for each model weren't immediately transparent. Users couldn't easily predict what a given action would cost.

Cursor has iterated since — clearer per-action cost displays, better dashboards, more granular budget controls. The credit-pool model is the current state and reflects the underlying economics. Whether you like it or not, it's what you're working with.

How Credits Work

A credit is roughly equivalent to a dollar of usage at standard rates. Your $20/month Pro plan includes $20 worth of credits. Each AI action consumes credits based on:

- **Which model** the action used. Opus 4.7 costs much more than DeepSeek V4 Flash.
- **How many tokens** were consumed in input and output. Long prompts and long responses cost more.
- **Whether thinking mode was on** (where applicable). Reasoning chains generate additional tokens.
- **How many turns** the agent took (for Agent and Background Agent). Multi-turn loops compound.

The dashboard shows your credit balance in real time and breaks down recent consumption by action and model. Settings → Account → Usage gives you the per-action ledger.

When you run out of credits, three things can happen:

Hard cap (default). Frontier model access stops. Tab and Cursor's fast model still work; everything else is gated until next month or until you upgrade.

Pay-as-you-go. You opt into usage-based billing beyond your included credits. Credits past the included amount go to your card at the same per-token rates. No discount for going over; usage costs the same per-unit whether it's "included" or "overage."

Budget cap with hard stop. You set a maximum monthly spend. Cursor warns you as you approach the cap and stops you at it, whether the cap sits above or below the included $20. The safest default for users who don't want surprise charges.

What Each Action Roughly Costs

Some rough numbers for credit consumption as of early 2026 (specific rates change; treat these as orders of magnitude):

Tab autocomplete on Cursor's fast model: essentially free. Hundreds of accepts per day costs cents.

Cmd+K with a small selection on DeepSeek V4 Flash: a fraction of a cent. Tens of invocations per day comes to tens of cents over a month.

Cmd+K with a larger selection on GPT-5.4: 1–3 cents per invocation. Tens per day adds up to a dollar or two per month.

Chat in Ask mode on Claude Opus 4.7: 5–15 cents per round, depending on context size.

Manual edit on Opus 4.7 with several files attached: 10–30 cents per round.

Foreground Agent on Opus 4.7 for a substantive task: $1–5 per task.

Background Agent on Opus 4.7 for a ticket-shaped task: $2–10 per task.

Multitask with three Opus subagents: $6–20 per run.

The numbers add up. A heavy user invoking Agent on Opus a dozen times a day burns through a $20 plan's credits in three days. The same user on DeepSeek V4 Pro for the same workload lasts the month.

The "$20 in Three Days" Trap

The most common pain pattern for new Cursor Pro users:

1. Sign up for Pro. Get $20 of credits.
2. Use Cursor for a day. Default model is Auto, but most chat invocations land on Opus.
3. Get excited about Agent mode. Send several Agent runs.
4. Day two: still feels great. Don't notice the credit balance dropping.
5. Day three: notification that credits are at 20%. Confused — "I just signed up."

6. Day four: out of credits. Stuck on Tab and the fast model for the rest of the month.
7. Either upgrade to Pro+ in frustration or write off Cursor as too expensive.

The trap is real; the fix is workflow discipline. The same user, with picker discipline (DeepSeek for routine, Opus for hard, Agent only for tasks worth running), can comfortably stay within $20/month for full daily use.

Forecasting Your Monthly Burn

A rough framework for estimating what you'll spend in a month:

Tab usage: essentially free. Don't budget for it.

Cmd+K usage: varies by model. Assume 10–50 cents per day if you use DeepSeek for most invocations and Opus for the occasional hard one.

Chat exploration: depends heavily on volume. A user who runs 5–10 substantive Ask sessions per day on a mix of models lands at $1–3 per day, or $20–60/month.

Manual edits: typically a few cents to a few dimes per round. A heavy user lands at $1–2 per day, or $20–40/month.

Foreground Agent: the variable that swings the budget most. Each run costs $1–5. Five runs per day is $5–25/day, or $100–500/month — well past Pro+ territory.

Background Agent: similar per-run cost to foreground. Volume tends lower (you don't kick off ten Background Agents in a day) but unit costs match.

Multitask: expensive per use, used selectively. A weekly run at $10–20 is reasonable; a daily Multitask habit is $200+/month.

For typical use:

- Light users: $5–10/month of actual consumption. Pro is fine; budget cap at $25 is plenty.
- Moderate users: $15–25/month. Pro fits; stay close to the included credits.
- Heavy users: $40–80/month. Pro+ fits ($60 included).

- Power users with daily Agent and Multitask: $100–300/month. Ultra fits ($200 included), or pay-as-you-go on Pro+.

If your forecast doesn't match your actual usage after a month, recalibrate.

Hard Caps and Budget Alerts

The single most useful setting for new Cursor users: set a hard budget cap.

In Settings → Billing → Spending Limits, you can configure:

Monthly cap. Maximum total credit consumption per month. When you hit it, Cursor stops; no overage charges.

Alert thresholds. Get notified at 50%, 75%, and 90% of the cap. The 75% notification is the actionable one — it tells you to rein in for the rest of the month.

Per-action cap. Maximum cost per single action (e.g., per Agent run). Prevents a single runaway agent from draining your pool.

Email notifications. Ensures you get alerts even when you're not actively using Cursor.

The book's recommendation: set the monthly cap at 1.5x your Pro plan's included credits during your first month. So $30 cap on a Pro plan with $20 included. The 50% headroom protects against surprise overages without blocking real work. After a month of usage data, adjust the cap to match your actual pattern.

What to Do When You Hit the Cap

You're at 95% credits, mid-month, and still need to ship work. Options:

Switch to cheaper models. DeepSeek V4 Pro and Flash run dramatically cheaper than Opus. Switching the default for the rest of the month stretches the remaining credits substantially.

Lean harder on Tab and Cmd+K. Tab is free. Cmd+K on the fast model is nearly free. Work that fits this size of edit doesn't tap the credit pool.

Avoid Agent and Background Agent. These are the biggest credit consumers. Manual mode is much cheaper for the same kind of work — just with you in the loop more.

Use BYOK temporarily. Chapter 20 covers how to point Cursor at your own API key and bypass the credit pool entirely. A useful stopgap when credits are exhausted but work continues.

Upgrade to Pro+ or pay-as-you-go. If you hit the cap consistently, the plan is wrong, not the workflow. Upgrade.

Ride out the month. Sometimes the right answer is "use Tab and the cheap models for two weeks and let the next billing cycle reset things." Fine for hobby use.

The mistake is panicking and mass-upgrading to Ultra without diagnosing what consumed the credits. Diagnose first. The dashboard tells you whether your spend went to Agent runs, chat sessions, or specific models. Address the cause.

Per-Project and Per-User Visibility

For Teams and Enterprise plans, the credit pool is typically shared across the organization. Admins see per-user breakdowns in the team dashboard. Common admin patterns:

Watch for outliers. One engineer consuming 5x the team's average is worth a conversation. Sometimes there's a legitimate reason (big refactor work), sometimes a pattern fix (they're using Opus for everything when DeepSeek would do).

Track per-project usage. If a particular project consumes disproportionate credits, it might benefit from rules tuning, BYOK setup, or scope adjustments.

Set alerts before hitting the team cap. A team running out of credits on day 25 is disruptive in a way that one user running out of personal credits isn't. Tighter cap discipline matters more for shared pools.

For Pro and Pro+ users, visibility is per-user and personal. The dashboard gives you enough to diagnose your own consumption patterns.

Cost Discipline Patterns

Three high-leverage habits:

The "DeepSeek by default, Opus by exception" rule. Set DeepSeek V4 Pro as your default. Switch to Opus when the task genuinely needs it. Most users get an order of magnitude in cost reduction with marginal quality loss, because most tasks don't need Opus.

The Manual-vs-Agent ratio. Manual mode costs less than Agent mode. If your workflow has shifted toward Agent for everything, deliberately push some tasks back to Manual. Credit savings are substantial, and review fidelity goes up too.

The pre-prompt pause. Before sending a request, take a half-second to consider whether you need the request at all. "Could I just write this myself?" "Could Tab do this if I started typing?" "Is this task actually scoped enough for Agent, or am I delegating something I haven't thought through?" The pause is small; the credit savings compound.

Per-Model Comparison Costs

A rough relative cost table for the same task across models, normalized to Opus 4.7 as 1.0:

Model	Relative cost	Notes
Claude Opus 4.7	1.0×	Premium reasoning
GPT-5.4	0.7–1.0×	Comparable, slightly cheaper
Gemini 3.1 Pro	0.6–0.9×	Cheaper, especially on long context
DeepSeek V4 Pro	0.1–0.15×	The cost-quality leader
DeepSeek V4 Flash	0.05×	Cheapest frontier-tier option
Cursor fast model (Tab)	~0×	Effectively free on paid plans

The DeepSeek line is the one most users underuse. The 10x cost gap to Opus translates directly into months of useful work for anyone who internalizes the picker discipline.

Chapter 20: Bring Your Own API

Cursor's credit pool is the default way you pay for model usage. Bring Your Own API (BYOK) is the alternative: configure Cursor to talk to your own API endpoint — Anthropic, OpenAI, OpenRouter, Together AI, Groq, your self-hosted DeepSeek, your company's internal LLM gateway — and Cursor sends requests there instead of through its credit pool.

BYOK is the right tool for some users and the wrong tool for many. This chapter covers when it makes sense, when it doesn't, and the mechanics of setting it up.

What BYOK Does

When you configure a custom API key in Cursor settings, requests for the configured model go directly to your endpoint. The model provider (Anthropic, OpenAI, etc.) bills you at their per-token rates, and Cursor doesn't deduct credits from your pool for those requests. Cursor still takes a small per-request fee from your subscription for the harness around the model — context retrieval, the diff interface, the agent loop infrastructure — but the model-token cost moves to your provider's bill.

The configuration is per-model, not all-or-nothing. You can run BYOK for Claude Opus while staying on Cursor's credit pool for GPT and DeepSeek, or vice versa. Mix and match by where you have favorable provider economics.

When BYOK Makes Sense

Three situations where BYOK is the right choice:

You have substantial enterprise credits with a model provider. If your organization has a $50K Anthropic contract with negotiated rates, those credits aren't accessible through Cursor's credit pool — Cursor buys at its own rate. BYOK lets you point Cursor at your enterprise key and burn down credits you've already paid for. Savings on a serious enterprise contract can be substantial.

You're self-hosting models. Teams running DeepSeek (or Llama, Mistral, etc.) on their own GPUs via vLLM, BentoML, or Ollama can point Cursor at their inference endpoint. Inference is essentially free at the per-request level (you've already paid for the GPUs), and credit-pool cost drops to zero. Particularly useful for organizations with strict data-locality requirements where hosted models aren't an option.

You're using cheap-tier providers like Together AI, Groq, or OpenRouter. Some providers price aggressively on the same models — DeepSeek hosted on Together at lower rates than Cursor's credit-pool pricing for DeepSeek, Llama variants on Groq at extremely low latency, OpenRouter as a price-shopping aggregator. BYOK lets you tap these without leaving Cursor's editor experience.

You want unlimited use of a specific model and have an enterprise key for it. If your company's Anthropic key has effectively unlimited rate limits for internal use, BYOK lets you run Opus 4.7 in Cursor without Cursor's credit cap. Marginal cost is low; the productivity unlock is real.

When BYOK Is the Wrong Tool

Five situations where BYOK is more friction than benefit:

You're a solo developer on Pro. Cursor's credit pool is reasonably priced for solo use. BYOK setup overhead rarely pays off on personal subscriptions. You'd save dollars at the cost of ongoing configuration maintenance.

You don't have a separate API key for the model. If you'd sign up for an Anthropic API account just to BYOK, weigh the new account and billing setup against simply paying Cursor. Convenience usually wins for individuals.

Your usage is low. If your monthly Cursor consumption is $10–20 in credits, BYOK saves very little. The break-even on individual plans is usually $50+/month in actual consumption.

You want Cursor's specific model offerings without provider direct accounts. Cursor's picker is curated — the four frontier models plus the Cursor-specific fast model. If you want exactly that mix without managing multiple provider accounts, the credit pool is a feature.

Your team has consolidated billing through Cursor. Enterprise teams often prefer a single Cursor invoice over a fan-out of provider invoices. BYOK fragments billing in ways finance teams don't always appreciate.

Setting Up BYOK

The configuration is in Settings → Models → API Keys. The flow varies slightly by provider:

Anthropic (for Claude models). Settings → Models → Anthropic → "Use my own API key." Paste your Anthropic API key (from console.anthropic.com). Cursor validates the key and offers a per-model toggle — you can use BYOK for Claude Opus while staying on credit pool for Sonnet, for example.

OpenAI (for GPT models). Same pattern. Settings → Models → OpenAI → "Use my own API key." Paste from platform.openai.com.

Custom OpenAI-compatible endpoints. The path for Together AI, Groq, OpenRouter, self-hosted vLLM/BentoML/Ollama, and any other provider exposing an OpenAI-compatible REST API. Settings → Models → Custom Endpoint → enter the base URL, the API key, and the model identifiers Cursor should expose. Cursor sends OpenAI-format requests to your endpoint.

Self-hosted DeepSeek via vLLM. As an example: spin up vLLM with the DeepSeek V4 model loaded, expose port 8000. In Cursor settings, add a custom endpoint with base URL `http://your-server:8000/v1/`, no API key (or a dummy one if vLLM requires it), and model name `deepseek-v4-pro`. Cursor routes requests for that model to your vLLM instance. The credit pool isn't touched.

Per-model routing. Once BYOK keys are added, the picker shows both BYOK and credit-pool options for the relevant models. Switch per request — BYOK Opus for heavy work, credit-pool DeepSeek for light work, all in the same chat session.

A Worked Example: Routing Everything to Self-Hosted DeepSeek

A scenario from a real team:

Goal: A 10-person engineering team running their own DeepSeek V4 deployment on internal GPUs. Wants to use Cursor's editor without paying Cursor's credit-pool rates for model usage.

Setup:

1. Deploy DeepSeek V4 Pro via vLLM on internal GPUs. Expose an OpenAI-compatible API at `https://llm.internal/v1/`.
2. In Cursor (Teams plan), each engineer goes to Settings → Models → Custom Endpoint, adds the internal URL, and configures the model name `deepseek-v4-pro`.
3. The Cursor admin pushes a shared MCP and rules configuration that includes a project rule: "Default model is deepseek-v4-pro via the internal endpoint."
4. Engineers use Cursor normally; chat, Cmd+K, Manual, and Agent all route to the internal DeepSeek deployment.

Result: The team's Cursor subscription costs are the per-seat editor fee ($40/user/month for Teams), not model usage. Model usage is amortized across GPU infrastructure they were already running. Net cost is meaningfully lower than the credit-pool equivalent for heavy users.

Caveats:

- Internal infrastructure has to be reliable. If the DeepSeek deployment goes down, Cursor can fall back to the credit pool, but the team loses productivity until the infra is back.
- Rate limits and concurrency on the internal deployment have to be sized for team usage. A small GPU cluster serving ten heavy Cursor users will queue requests if undersized.
- BYOK doesn't enable Privacy mode by itself. Code still flows through Cursor's infrastructure for context retrieval and the agent loop; only the model-call portion stays internal. Chapter 21 covers Privacy mode separately.

When to Mix BYOK and Credit Pool

The most common BYOK pattern isn't "all BYOK" but "BYOK for the expensive model, credit pool for everything else."

A typical configuration:

- **Claude Opus 4.7:** BYOK against Anthropic enterprise key. The expensive model has the most savings from direct provider rates.
- **GPT-5.4:** Credit pool. Convenience over savings; usage is bounded.
- **DeepSeek V4 Pro:** Credit pool. Already cheap enough.
- **Gemini 3.1 Pro:** Credit pool. Same.

This mix gives you the best of both worlds — the expensive model's cost is controlled, and the rest of the picker stays simple. The downside is configuration complexity (one provider account, one BYOK setup), but it's a one-time cost.

BYOK and Privacy

BYOK changes some of the data flow but not all of it:

With BYOK against your own provider key. The model-call goes directly to the provider — Anthropic, OpenAI, etc. — using your account. Cursor's servers don't see the model's input/output for those calls. They still see the surrounding context (file paths, project structure, retrieval queries) for the editor harness.

With BYOK against a self-hosted endpoint. Even more isolation. The model invocation never leaves your network. Cursor's servers only see whatever metadata and harness telemetry Cursor itself collects.

This is partial privacy, not zero-data-retention. Privacy mode (Chapter 21) is a separate setting that controls what Cursor's own infrastructure stores. BYOK and Privacy mode are complementary; the strongest privacy posture combines both.

For regulated workloads, the typical configuration is:

- BYOK against a self-hosted model deployment in your environment.
- Privacy mode on at the Cursor account level.
- Enterprise plan with custom DPA and data-handling terms.

Together, these minimize what touches Cursor infrastructure. Some workloads still won't fit (tasks requiring shared MCP integrations, Background Agents needing cloud sandbox infrastructure), but for editor-only use, the posture is strong.

BYOK Limitations

Some things BYOK doesn't change:

Background Agents still run on Cursor's cloud. The sandbox infrastructure is Cursor's. If your security model requires all agent execution stay in your environment, Background Agents aren't compatible regardless of BYOK.

Indexing still happens on Cursor's servers. The codebase index — embeddings, retrieval, the Merkle tree — runs on Cursor's infrastructure. BYOK doesn't move indexing to your servers. Self-hosted indexing is an Enterprise contract feature.

MCP-integrated workflows route through Cursor. When Agent calls an MCP server, the MCP infrastructure is Cursor's wrapping around your configured tools. Data flow goes through Cursor's harness even if the model itself is BYOK.

Cursor's per-seat subscription still applies. BYOK saves on model costs, not editor cost. A Teams seat is $40/user/month regardless.

Some features may be premium-only. Cursor occasionally restricts new features to certain plans or requires credit-pool models. BYOK doesn't bypass these gates.

Picking a BYOK Provider

For users who decide BYOK is the right path, the question becomes which provider:

Anthropic direct. The only path for Claude models. Sign up at console.anthropic.com, generate a key, plug into Cursor. Pricing matches Anthropic's published rates; enterprise contracts get negotiated rates.

OpenAI direct. The only path for GPT models. Sign up at platform.openai.com, generate a key, plug in.

Together AI. Hosts open-weight models (DeepSeek, Llama, Mistral, etc.) at competitive rates. Good fit for users who want open-weight model access at lower-than-Cursor pricing without self-hosting.

Groq. Specializes in low-latency inference. Limited model selection but extremely fast for what they serve. Good when latency matters more than selection.

OpenRouter. A meta-provider that routes to many backends. Good for price-shopping or users who want a single API for many models. Slightly higher latency from the routing layer.

Self-hosted (vLLM, BentoML, Ollama). The most flexibility, the most operational overhead. Right for teams already running model infrastructure for other reasons; usually wrong for teams who'd stand it up just for Cursor.

The book's recommendation for mid-sized teams: start with credit pool, add BYOK against Anthropic enterprise key when usage scales past $200/month for Opus specifically, add self-hosted DeepSeek when you reach scale where the GPU economics make sense (typically 20+ heavy users).

Routing Without Owning the Pool

BYOK controls where model invocations go. Privacy mode controls what Cursor retains about those invocations. They're complementary rather than overlapping. Chapter 21 covers Privacy mode: the threat model, what it does and doesn't protect against, and how to choose it for your context.

Chapter 21: Privacy Mode

Privacy mode is the setting most users either turn on without thinking too hard or turn off and never reconsider. Both are mistakes. The setting matters, the threat model is specific, and getting it right depends on the code you're working on. This chapter is about understanding what Privacy mode does, what it doesn't, and how to make the call deliberately.

What Privacy Mode Is

Privacy mode is a per-account setting that changes how Cursor's infrastructure handles your code. The high-level distinction:

Privacy mode off (default). Cursor processes your code through its normal infrastructure. Plaintext code goes to Cursor's servers for embedding generation, model invocation, and various other server-side operations. Cursor retains some of this data per its standard data-handling policy — primarily for product improvement, debugging, and service operation. The retention policy is documented at cursor.com/privacy.

Privacy mode on. Zero-data-retention infrastructure. Plaintext code is processed in transit but not retained. Cursor receives the code, does what it needs to do (generate an embedding, send a model call, return a response), and discards the plaintext. The vector index is still stored (retrieval needs it), but the original plaintext behind the vectors isn't kept.

The setting lives at Settings → Privacy → Privacy Mode. Toggling takes effect immediately for new requests. Existing data isn't retroactively purged when you turn the setting on; it's a forward-looking change.

For Enterprise plans, Privacy mode can be enforced org-wide as a policy, not a per-user choice. Admins decide whether engineers can override it.

What Privacy Mode Protects

The specific protections under Privacy mode:

No plaintext retention. Cursor's servers don't store your code beyond the request that generated it. If a request fails and a debugger goes looking at logs, the code isn't there.

No training data use. Cursor doesn't use your code to train models. This holds regardless of Privacy mode for paid plans, but Privacy mode adds the structural guarantee that the code couldn't be used for training even if the policy changed — because it isn't stored to be used.

Reduced telemetry. Some telemetry collected by default (request payloads, error contexts) is reduced or omitted under Privacy mode. The exact specifics are documented at cursor.com/privacy.

Compliance posture. Privacy mode is the primary control for regulated workloads. For SOC 2-aligned, HIPAA-aligned, FedRAMP-adjacent workflows, Privacy mode is a starting requirement (necessary but not always sufficient).

What Privacy Mode Doesn't Protect

The specific things Privacy mode is not:

It is not "Cursor doesn't see your code." Cursor's servers process the code in transit. The model providers (Anthropic, OpenAI, DeepSeek, Google) also see the code sent for inference. Privacy mode controls retention, not visibility-during-processing.

It does not change the provider's data handling. When Cursor sends a request to Anthropic, what Anthropic does with that data is governed by your relationship with Cursor (which has data-processing agreements with each provider), not by your Cursor Privacy mode setting. Most providers also offer their own zero-data-retention modes for enterprise customers.

It is not encryption-at-rest of your code. Code on your laptop is your laptop's responsibility. Privacy mode covers Cursor-side handling; local encryption is unaffected.

It is not legal cover. "I had Privacy mode on" doesn't satisfy your organization's compliance team if their requirements run more specific. Privacy mode is a technical setting; compliance posture is a contractual matter.

It is not a substitute for `.cursorignore`. Files you don't want Cursor to see at all — credentials, secrets, personally identifiable data — belong in `.cursorignore`. Privacy mode reduces retention; `.cursorignore` reduces transmission. Use both for sensitive content.

When to Turn Privacy Mode On

For most users, Privacy mode should be on. The default at sign-up is off; the recommendation is to turn it on at first launch unless you have a specific reason not to. The latency impact is negligible. The privacy posture is meaningfully better. There's no daily-workflow downside.

Specifically, Privacy mode should be on when:

You're working on proprietary code. Even if your employer hasn't explicitly required it, treating your employer's IP with care is just professional. Privacy mode is the structural mechanism for that care.

You're working under a contractual or regulatory regime. Customer contracts that require data isolation, regulatory frameworks (HIPAA, SOC 2, FedRAMP-adjacent), industry requirements that touch your codebase. Privacy mode is the start of compliance posture; it doesn't finish it but it's a necessary piece.

You're working with secrets, credentials, or PII in code. Even with `.cursorignore` exclusions, defense in depth matters. Privacy mode is one layer; the ignore file is another.

You're working in security-critical code. Authentication, authorization, payment processing, data integrity. Code where the cost of accidental disclosure is high.

When Privacy Mode Off Is Acceptable

There are situations where Privacy mode being off is fine:

You're working on open-source code. If the code is already public, the marginal privacy gain from Privacy mode is zero. Turning it off adds a small amount of telemetry to Cursor's product-improvement signals; it costs you nothing.

You're learning or experimenting on throwaway code. Tutorial projects, sample apps, code with no IP value. Privacy mode is fine but unnecessary.

You're explicitly opting in to feature development. Some early-access or beta features that depend on retained data don't work under Privacy mode. If you're a power user opted into experimental features, the off setting may be required.

For most users, none of these apply daily, so the practical default is "Privacy mode on, always."

The Enterprise Overlay

Enterprise plans add several features on top of Privacy mode:

Org-wide enforcement. The admin sets Privacy mode as a policy; individual engineers can't override.

Custom DPAs. Data Processing Agreements with Cursor that specify exactly what data is processed, retained, and for how long. Enterprise contracts often include custom language for the customer's regulatory regime.

Data residency. Some Enterprise contracts include geographic data-residency commitments — your data is processed in specific regions. Default Cursor processing is multi-region.

Audit logs. Comprehensive logging of which engineer accessed which repository and made which AI invocations. Used for compliance review and incident investigation.

Self-hosted indexing. For the strongest privacy posture, Enterprise contracts can include self-hosted indexing where the codebase index lives on the customer's infrastructure rather than Cursor's. A meaningful operational lift on the customer side.

Repository, model, and MCP blocklists. Admins specify that certain repositories never get indexed, certain models can't be called, certain MCP servers can't be installed. These bound the surface area where AI invocations can happen.

For organizations with formal security review processes, the Enterprise overlay is what makes Cursor approvable. Without it, Privacy mode alone is often insufficient for the security team's review.

The Threat Model: What's Actually At Risk

To choose Privacy mode well, it helps to think clearly about what could go wrong without it.

Risk: data breach at Cursor's infrastructure. A breach exposes whatever Cursor was retaining. Without Privacy mode, that includes recent plaintext code from your projects. With Privacy mode, it's only the vector index and metadata.

Mitigation effectiveness: High. Privacy mode genuinely reduces the impact of a breach.

Risk: legal compulsion. A subpoena or government request demands Cursor turn over data. Whatever Cursor has, it has to turn over. Privacy mode means there's less to turn over.

Mitigation effectiveness: High for retention-based requests. Doesn't help for "log future requests for this user," a different legal mechanism.

Risk: insider threat at Cursor. A Cursor employee accesses customer data they shouldn't. Privacy mode reduces what's available to access.

Mitigation effectiveness: Moderate. Cursor's internal access controls matter at least as much.

Risk: Cursor bug or misconfiguration. A bug causes Cursor to mishandle data. Privacy mode reduces what's affected if the bug sits in retention pipelines.

Mitigation effectiveness: Moderate. Doesn't help if the bug sits in transmission or processing.

Risk: model provider data handling. Anthropic/OpenAI/DeepSeek do something with your code beyond what you expected. Cursor's Privacy mode doesn't control this — provider relationships do.

Mitigation effectiveness: Zero. Requires provider-side controls (their zero-data-retention modes, BYOK against your own enterprise keys with custom data terms, etc.).

The threat model isn't paranoid — these are real risks at scale. They're also bounded. Most are addressed structurally by Privacy mode plus reasonable infrastructure hygiene. For organizations where the residual risk is unacceptable, the Enterprise overlay and BYOK against self-hosted models close more of the gap.

Combining Privacy Mode with Other Controls

The strongest privacy posture in Cursor combines multiple controls:

Privacy mode on. Reduces Cursor-side retention.

`.cursorignore` for sensitive files. Excludes files from being sent at all.

BYOK against your own keys. Routes model invocations through your provider relationship rather than Cursor's. Combined with provider-side zero-data-retention, the model-call path doesn't retain your code anywhere.

Self-hosted models for the most sensitive workloads. Inference happens entirely in your environment.

Enterprise contractual terms. DPAs, custom data handling, audit logs, geographic residency.

Repository scoping at the org level. Admins block Cursor from being used on certain repos at all.

For most teams, "Privacy mode on" is the only one they need. For regulated teams, several or all apply. Choose by what your context actually requires; layered controls cost overhead and shouldn't be applied for posture without reason.

Privacy and the AI Layer's Limits

A subtle point worth raising: even with maximum privacy posture, AI-assisted development inherently exposes patterns to AI infrastructure. The model provider sees what tasks you ask about. Even if no plaintext is retained, usage patterns — which APIs you call, which problems you solve, what your team is working on this quarter — are visible during processing to whoever runs the inference.

For most organizations, that's fine. The model providers have strong reputational and contractual incentives to handle this responsibly. For some organizations — defense, intelligence, certain financial contexts — even this level of pattern exposure is too much. Those organizations either don't use AI editors, use them only on non-sensitive code, or self-host every layer of the stack to keep pattern data local.

This isn't a Cursor-specific issue; it applies to every hosted AI tool. Privacy mode addresses retention; pattern exposure during processing is a different question no current product fully solves.

Threat Modeling Cursor for Your Security Team

If you're advocating for Cursor in an organization with a formal security review, the threat-model document your security team will want includes:

- **Data flow diagram.** What data goes to Cursor's servers, what stays local, what flows to model providers.
- **Retention policy.** What Cursor stores, for how long, under what circumstances.
- **Privacy mode behavior.** How Privacy mode changes the retention picture.
- **Sub-processor list.** Which model providers Cursor uses, which other vendors are in the pipeline.
- **Incident response.** What Cursor's process is for handling a security incident.
- **Compliance attestations.** SOC 2 Type 2 report, GDPR posture, any other relevant attestations.
- **Customer references.** Other organizations in your industry using Cursor under similar requirements.

Cursor's enterprise team provides most of this on request. The book's recommendation is to engage Cursor's enterprise sales early in the security review — they've been through dozens of these conversations and can provide the materials your security team will eventually ask for, often before the questions arrive.

Part V closes here. The model layer — picker, credits, BYOK, Privacy mode — is configured. Part VI moves into the ecosystem that extends Cursor's reach to external systems and shared team workflows: MCP, BugBot, hooks and slash commands and Skills, and the terminal/debugger integration. Chapter 22 starts with MCP, the most consequential ecosystem feature of 2026.

Part VI: The Cursor Ecosystem

"A tool's reach is the union of its plug points and the things plugged into them."

— A working principle

Chapter 22: MCP — Model Context Protocol

MCP — the Model Context Protocol — is the most consequential ecosystem development in Cursor's 2026. It's the standard that lets external systems plug into the editor's AI features as first-class tools. With MCP installed and configured, Agent mode can read your Linear tickets, query your Sentry errors, run queries against your dev Postgres, post to Slack, pull context from Notion, and dozens of other things — all without leaving the editor.

Before MCP, integrations like these were custom one-offs. Now they're a protocol with hundreds of servers and a curated catalog inside Cursor. This chapter covers what MCP is, how to use it, how to build your own, and the discipline that keeps an MCP-enabled workflow from sprawling into a mess.

What MCP Is

MCP is an open protocol developed by Anthropic in 2024 and broadly adopted across the AI tool ecosystem in 2025–2026. It specifies how an AI client (Cursor, Claude Code, Claude Desktop, others) talks to a context provider (an "MCP server") over a small set of operations:

- **List tools.** What capabilities does this server offer?
- **Call tool.** Invoke a capability with arguments. Get back a result.
- **List resources.** What context can this server provide?
- **Read resource.** Fetch a piece of context.
- **List prompts.** What templated prompts does this server make available?

An MCP server is a small program — usually written in TypeScript or Python with a thin SDK — that implements these operations against some external system. A Linear MCP server wraps the Linear API. A Postgres MCP server wraps a database connection. A GitHub MCP server wraps the GitHub API. The protocol stays the same regardless of what's behind it; the editor doesn't need to know the specifics.

When you install an MCP server in Cursor, its tools become available to Agent mode (and to chat with tool use enabled). The model sees them in its tool list and calls them autonomously when the task warrants. From your perspective, you ask Agent to "fix issue #1234," Agent fetches the issue from Linear, reads the description, makes the fix, and updates the issue's status — all in one autonomous run.

The Three Transports

MCP servers communicate with Cursor over one of three transport mechanisms:

stdio. The MCP server runs as a child process of Cursor. Communication happens over stdin/stdout. The lowest-overhead transport for local servers — Postgres connections to local databases, filesystem operations on local paths, anything that doesn't need to live elsewhere.

SSE (Server-Sent Events). A legacy HTTP transport — Cursor connects to an MCP HTTP server over a long-lived event stream. SSE was deprecated in the MCP spec in March 2025 in favor of Streamable HTTP, with deprecation deadlines arriving mid-2026. Existing SSE servers still work for now, but new servers should use Streamable HTTP.

Streamable HTTP. The current standard for production-grade remote servers. Supports better authentication patterns, more robust reconnection, OAuth 2.1, and per-tool authorization. The default for new public MCP servers in 2026 and the recommended choice for anything beyond local experimentation.

The transport choice affects setup complexity and operational characteristics, but the protocol on top is the same. From the model's perspective, it doesn't matter how the MCP server is reached — it calls tools and gets results.

Global vs. Per-Project Scope

MCP servers can be installed at two levels:

Global. Configured at the user level, available in every project. Right for tools you use everywhere — your personal notes app, a sequential-thinking server, a generic web-fetch tool.

Per-project. Configured in the project's `.cursor/mcp.json` file (checked into version control, shared with the team). Right for tools specific to this project — the project's database, its issue tracker, its cloud provider integration.

The split matters because per-project configuration ships through the repo, so the team gets the same MCP setup automatically. New engineers cloning the project get the project's MCP servers without manual setup. Configuration updates ship as code changes through normal PRs.

The recommendation: keep your personal toolkit at the global level (a few servers you use everywhere) and put project-specific servers in `.cursor/mcp.json`. The split keeps the global setup small and the project configuration shareable.

One-Click Installs from the Catalog

Cursor 2026 ships with a curated MCP catalog accessible from the command palette ("Install MCP Server"). The catalog includes vetted servers for common integrations:

- **GitHub** — read repos, manage PRs, run actions, comment on issues
- **Linear** — read tickets, update status, add comments, link PRs to issues
- **Sentry** — query errors, fetch traces, create alerts
- **Postgres** — connect to databases, run queries, inspect schemas
- **Notion** — read pages, search content, post updates
- **Stripe** — query customers, payments, subscriptions (read-only by default)
- **Slack** — read messages, post notifications
- **Jira** — read tickets, update status, add comments (for the Atlassian-aligned shops)
- **Filesystem** — extended filesystem operations beyond the editor's defaults
- **Sequential Thinking** — a thinking-aid server that helps the model decompose hard problems
- **Web Fetch** — generic HTTP fetching beyond the editor's built-in web access

Install is one click. The catalog handles the configuration boilerplate — you authenticate once via OAuth or paste an API key, and the server is configured.

Updates ship via the catalog; you don't manage server versions manually for catalog installs.

For servers outside the catalog (custom servers, internal company servers, third-party servers not yet curated), configure manually via `.cursor/mcp.json` — covered later in this chapter.

The 40-Tool Ceiling

A practical limit worth knowing: Cursor caps the number of tools that can be active at once at around 40. The cap exists because the model's tool-selection performance degrades when too many tools are visible — more choices, more confusion, lower per-tool accuracy.

If you install many MCP servers and exceed the cap, Cursor warns you and requires you to disable some servers (or specific tools within servers) before activating new ones. The cap shapes how to think about MCP installation:

Don't install everything. The temptation, with one-click installs, is to add every catalog server "just in case." Resist. Every installed tool is a tool the model weighs on every invocation; unused tools dilute the signal.

Install per-project rather than global. Project-scoped tools only count toward the cap when you're in that project. A different project can have a different mix. Keeping the global mix small leaves headroom for per-project tools.

Disable rather than uninstall. Cursor lets you disable individual tools within a server. If a server has 15 tools and you use 3, disable the other 12. Disabled tools don't count against the cap.

Periodic cleanup. Every quarter or so, audit your installed servers. Anything unused in 30 days — disable or uninstall. The accumulation pattern is real, and the model's behavior degrades as the tool list grows.

The 40-tool ceiling sounds restrictive but in practice is generous. Most teams use 10–20 tools across all their MCP servers. Hitting the cap signals over-installation.

Calling MCP Tools from Agent

When MCP servers are installed, Agent mode discovers their tools automatically and calls them as needed. You don't typically need to mention specific tools in your Agent prompt — the model sees the available tools, reads their descriptions, and chooses.

A concrete example. You've installed the Linear MCP server. You send Agent:

> "Fix issue ENG-1234. Read the issue description, implement the fix in our codebase, run the tests, and update the issue with a link to the PR when you open it."

Agent's loop:

1. Calls Linear's `get_issue` tool with `id: "ENG-1234"`. Reads the title, description, and assignee.
2. Reads the relevant files in the codebase based on the issue description.
3. Plans the fix.
4. Edits files, runs tests until they pass.
5. Pushes a branch and opens a PR.
6. Calls Linear's `update_issue` tool with the PR link in a comment, sets status to "In Review."
7. Reports back: "Done. PR #5678. Linear ticket updated."

The whole flow runs autonomously. You didn't tell Agent how to call Linear's API — the MCP server handles that. You didn't tell Agent how to format the comment — the model picks a sensible format based on what it sees in the project's history.

Calling MCP Tools Manually

You can also invoke MCP tools directly from chat without going through full Agent autonomy. The `@` syntax covers MCP servers; `@linear` followed by a description routes the request to Linear's tools.

Some patterns:

> "@linear show me my open issues from last week"

"@postgres run a query to count active users by signup month for the last six months"

"@sentry fetch the last 10 errors from the production deployment"

The model translates your natural-language request into the right MCP tool call, executes it, and shows you the result. Faster than full Agent for one-off queries where you want the data, not a full task execution.

Building Your Own MCP Server

The MCP SDKs (TypeScript, Python, others) make custom server development straightforward. A minimal server runs roughly 100 lines: define your tools, implement the handlers, run the SDK's server loop.

A skeleton of a custom MCP server in TypeScript:

```
import { Server } from "@modelcontextprotocol/sdk/server/index.js";
import { StdioServerTransport } from "@modelcontextprotocol/sdk/server/stdio.js";

const server = new Server(
  { name: "my-internal-tools", version: "1.0.0" },
  { capabilities: { tools: {} } }
);

server.setRequestHandler("tools/list", async () => ({
  tools: [
    {
      name: "deploy_to_staging",
      description: "Deploy the current branch to staging environment",
      inputSchema: {
        type: "object",
        properties: {
          branch: { type: "string", description: "Git branch to deploy" }
        },
        required: ["branch"]
      }
    }
  ]
}));

server.setRequestHandler("tools/call", async (request) => {
  if (request.params.name === "deploy_to_staging") {
    const branch = request.params.arguments.branch;
```

```
    const result = await deployToStaging(branch);
    return {
      content: [{ type: "text", text: result }]
    };
  }
  throw new Error(`Unknown tool: ${request.params.name}`);
});

const transport = new StdioServerTransport();
await server.connect(transport);
```

Configure Cursor to run this server in `.cursor/mcp.json`:

```
{
  "mcpServers": {
    "internal-tools": {
      "command": "node",
      "args": ["./tools/mcp-server.js"]
    }
  }
}
```

Cursor starts the server when needed, calls its tools as the model decides, and the deploy capability becomes available to Agent.

Custom MCP servers are how teams expose internal tooling to Cursor — deployment systems, internal admin tools, company-specific APIs, anything the team uses regularly that isn't in the catalog. The investment is small (a couple of hours per server) and the leverage compounds.

Walkthrough: Wiring Linear into Agent Mode

A concrete end-to-end example.

Step 1: Install the Linear MCP server from the catalog. Command palette → "Install MCP Server" → Linear → "Install." The catalog walks you through OAuth with Linear; you grant permissions for the workspace.

Step 2: Verify the install. Open chat, type `@linear`. The completion shows the available Linear tools: `get_issue`, `update_issue`, `create_issue`, `list_issues`, etc.

Step 3: Test with a simple query. "@linear list my open issues."

The model calls Linear's `list_issues` tool with appropriate filters and shows the result. If auth and connection are working, you'll see your actual ticket list.

Step 4: Use it in an Agent task. Switch to Agent mode. Send:

> "Look at my open Linear issues. Pick the smallest one with a clear scope and implement it. When done, update the issue with the PR link."

Agent calls `list_issues`, reads them, picks one, implements, pushes a PR, calls `update_issue`. The whole flow runs autonomously.

Step 5: Add to project rules. If your team has Linear conventions — specific status names, label requirements, format expectations for ticket comments — add them to `.cursor/rules/linear.mdc`:

```
---
description: Linear conventions for this project
globs: []
alwaysApply: true
---

# Linear Conventions

- Issue IDs are prefixed ENG-, INFRA-, or SEC-.
- When linking a PR to an issue, use the format "PR: <github-url>".
- Set status to "In Review" when opening a PR; "Done" when merged.
- Don't auto-close issues; leave that for the human reviewer.
```

The rule loads on every Agent run and shapes how the model uses the Linear MCP tools. The combination — MCP server for capability, project rule for behavior — is the pattern to internalize.

MCP Security

A concern that comes up: MCP servers do real things. The Linear server updates tickets. The Postgres server runs queries (potentially destructive). The deployment server deploys to staging. What stops the model from doing the wrong thing?

Several layers:

Tool descriptions. Each tool's description tells the model what it does and any constraints. Well-described tools include warnings ("This tool deletes the record. Confirm before calling.").

Per-tool authorization. Cursor can require explicit per-call confirmation for sensitive tools. The first time Agent wants to call a destructive tool, you're prompted. Grant for the session, deny, or always-require-prompt.

Read-only by default for sensitive servers. The catalog versions of Stripe, Postgres, and similar servers default to read-only. Write operations require additional configuration the user has to enable.

Per-project tool blocklists. Admins on Teams/Enterprise plans can blocklist tools they don't want available — "Postgres write operations are never allowed in Cursor for our team."

MCP server quality. Catalog-curated servers are vetted. Custom servers are your responsibility; if your custom MCP server has bugs or unsafe defaults, that's on you.

The book's recommendation: treat MCP tool capabilities like any privileged action. Be explicit about what you allow, audit what you've granted, and prefer read-only over write where the choice exists. Agent is good but not infallible; you don't want a confused agent loop running destructive operations because the tools were installed without thought.

MCP and Privacy

MCP servers communicate through Cursor. The data flow:

1. Cursor sends the model your prompt and the project context.
2. The model decides to call an MCP tool.
3. Cursor invokes the MCP server (over stdio, SSE, or HTTP) with the tool's arguments.
4. The MCP server does its work — possibly contacting external APIs.
5. The result returns through Cursor to the model.
6. The model continues with the result as context.

What this means for privacy:

MCP server inputs/outputs go through Cursor. Even for local stdio servers, the inputs and outputs to tool calls flow through Cursor's process. Privacy mode applies to retention.

MCP server external calls are visible to the external system. A Linear MCP server makes HTTP calls to Linear's API. Linear sees those calls. Cursor's Privacy mode doesn't change Linear's behavior.

Authentication tokens. API keys and OAuth tokens for MCP servers live in Cursor's settings (or per-project mcp.json). Watch what you check into version control — `.cursor/mcp.json` is shared with the team and shouldn't contain secrets. Use environment variables or secret management for credentials.

For sensitive workloads, this matters. MCP pulls external systems into the data flow, each with its own privacy posture. Audit each MCP server you install for what it talks to and how.

Chapter 23: BugBot — Automated PR Review

BugBot is Cursor's GitHub-integrated automated PR reviewer. It reads pull requests as they open, identifies issues, and — since the February 2026 graduation from beta — opens follow-up PRs to fix what it finds. The reported resolution rate on issues BugBot identifies is around 80%, high enough that BugBot meaningfully reduces review burden on teams that adopt it.

BugBot is also polarizing. Used well, it catches bugs human reviewers miss and frees senior engineers from low-value review cycles. Used poorly, it floods PRs with low-value comments, generates fix-PRs nobody wants, and erodes trust in automated review. This chapter is about getting it right.

What BugBot Is

BugBot is a separate product from the Cursor editor. It runs as a GitHub App; you install it on your repos and it watches for new PRs. When a PR opens (or updates), BugBot:

1. **Reads the PR diff** along with surrounding context from the codebase.
2. **Analyzes for issues** — bugs, regressions, missing edge case handling, style violations, security risks, performance concerns, missed test coverage, and other categories a careful reviewer would flag.
3. **Posts review comments** on the PR for issues it identifies, with explanations.
4. **For issues it can fix** (the new fixer capability), opens a follow-up PR with the proposed fix, branched from the original PR's branch.

The reviewer-only behavior was the original beta (2025). The fixer behavior is new with the February 2026 graduation; BugBot now opens its own PRs to fix the issues it identifies.

The model behind BugBot is a Cursor-tuned variant of frontier reasoning models. It has access to the codebase index, the project rules, and any MCP servers configured at the project level. From the PR's perspective, BugBot looks like any other reviewer — a GitHub username that reads, comments, and (now) opens fix PRs.

Pricing

BugBot is priced separately from the Cursor editor subscription:

BugBot Pro. Usage-based pricing on top of an existing Cursor Pro/Pro+/Ultra subscription. Pay per PR review and per fix opened. Flexible for individuals or small teams.

BugBot Teams. Included with Teams subscriptions, with usage limits. Most small-to-medium teams stay within the limits without overage.

BugBot Enterprise. Included with Enterprise contracts, custom SLAs, custom usage envelopes, and BugBot against private repos with strict security controls.

The decision to enable BugBot is partly cost (a high-PR-volume team pays non-trivial usage fees) and partly culture (does your team want automated review). The book's recommendation: pilot on a single repo for two weeks before committing to org-wide rollout.

What BugBot Catches Well

BugBot's strengths align with the things a careful but tireless human reviewer would catch:

Common bug patterns. Off-by-one errors, null-handling gaps, error swallowing, missing await on promises, type narrowing failures, condition inversions. The mechanical bugs human reviewers also catch but sometimes miss when tired.

Edge case omissions. "What happens with empty input?" "What if this query returns zero rows?" "What if this user has no profile?" — the unspoken edge cases that break production. BugBot flags these well because it doesn't get tired.

Security smells. SQL injection patterns, XSS vulnerabilities, missing authorization checks on routes, hardcoded secrets, unsafe deserialization. Not a substitute for a security review, but a strong second pass.

Test coverage gaps. PRs that change behavior without updating tests. PRs that add features without adding tests. PRs that delete tests "by accident." BugBot flags these reliably.

Style and convention violations. When project rules cover the convention, BugBot enforces them. Rules and BugBot work together — rules set the standard, BugBot catches violations.

Performance footguns. N+1 queries, blocking operations in async paths, unnecessary work in hot paths. The kind of performance issue that compiles fine and shows up under load.

What BugBot Catches Poorly

Equally important to know:

Architectural concerns. "This shouldn't be a service, it should be a library." "This abstraction is wrong." BugBot doesn't reliably catch design issues — it reads the diff, not the broader system. Senior reviewers still own these.

Business logic correctness. Whether a calculation is right depends on what it's supposed to calculate. BugBot can flag obvious math errors (`x + y` where the spec says `x - y`) but can't verify business correctness without the spec being available somehow (rules, comments, related issues).

Subtle race conditions and concurrency bugs. Concurrency reasoning is hard. BugBot catches the obvious patterns but misses the subtle ones. For concurrency-critical code, human review remains the primary line of defense.

Cross-PR consistency. "This PR contradicts an architectural decision we made in PR #1234." BugBot reviews each PR in isolation; it doesn't reliably catch inconsistency with prior decisions.

Strategic concerns. "This is the right code but the wrong feature." That's product judgment, not code judgment.

The honest framing: BugBot is a good first-pass reviewer that catches mechanical issues at high recall. It is not a senior engineer's review. Treating it as a complete review is the most common way teams get burned.

When BugBot Helps

BugBot is the right tool when:

Your team's PR volume exceeds review capacity. Engineers ship faster than reviewers can keep up. BugBot's first pass catches mechanical issues, freeing humans for the higher-value architectural and business-logic review.

Your team has consistent conventions BugBot can enforce. Project rules + BugBot is the most effective combination. Rules tell BugBot what your standards are; BugBot enforces them. Without rules, BugBot guesses at conventions.

You have flaky reviewers or inconsistent review quality. Some teams have one reliable reviewer and several drive-by ones. BugBot's consistency is a leveler — every PR gets at least the BugBot pass.

You're in a fast-moving startup where senior engineers are bottlenecked. Senior time is the constrained resource. BugBot lets seniors skip mechanical review and focus on architecture.

When BugBot Hurts

BugBot is the wrong tool when:

Your team lacks the discipline to triage BugBot's comments. BugBot's output needs review like any reviewer's. If comments accumulate without being addressed or dismissed, the PR queue becomes noise. Adopt BugBot only if your team will treat its comments seriously — addressing the real issues, dismissing the false positives.

Your codebase has weak or no project rules. Without rules, BugBot guesses at conventions and often guesses wrong. The result: comments that don't match your team's actual standards. Fix the rules first; then enable BugBot.

Your PR culture is already minimalist. Some teams ship small PRs reviewed quickly by a single human. BugBot's automated comments add friction without value — every PR now has comments to triage when previously it had none.

You don't want fix PRs. The fixer behavior opens new PRs to address issues. Some teams find this useful (issues get fixed faster); others find it noisy (extra PRs to review for the same code change). Disable the fixer if you prefer review-only behavior.

Configuring BugBot

After installing the GitHub App, BugBot's behavior is configured in the repo's `.bugbot/` directory (or equivalent — exact paths change as the product evolves):

Enabled paths and excluded paths. Scope BugBot to specific directories. Default is "everywhere"; many teams scope to backend code and exclude generated code, vendor directories, test fixtures.

Severity thresholds. BugBot categorizes issues by severity. Configure which severities get comments and which get auto-fix attempts. A common pattern: high-severity gets comments and a fix PR; medium-severity gets a comment only; low-severity is silenced.

Issue category toggles. Disable specific issue categories. Some teams turn off "missing JSDoc" comments because they don't enforce that. Some turn off performance suggestions because their performance optimization happens elsewhere.

Project rules integration. BugBot reads `.cursor/rules/` and uses them as standards for review. Rules that codify your team's conventions become things BugBot checks for.

Fix-PR settings. Whether BugBot opens fix PRs at all, which severities trigger fixes, branch naming conventions, PR description templates.

Configuration evolves as your team learns what's helpful and what's noise. The first month of BugBot adoption is iteration on the configuration, not just BugBot doing its work.

Triaging BugBot Comments

When BugBot comments on a PR, your options are:

Address the issue. The comment is right; fix the code, push the change, the comment resolves automatically.

Mark as won't fix. The comment is right but you've decided not to act on it (out of scope, deliberate choice, etc.). Mark resolved with a brief note.

Dismiss as wrong. The comment is wrong — false positive, missing context, hallucinated issue. Dismiss with a note explaining why. BugBot uses dismissals to calibrate over time.

Escalate. The comment surfaces a question that needs human judgment. Tag a human reviewer to make the call.

The discipline is treating BugBot's comments seriously enough to triage. Accumulating un-triaged BugBot comments is the failure pattern. Two responses every team should avoid:

Reflexively dismissing. "BugBot is annoying, mark all dismissed." That trains your team to ignore the bot, which is fine — but then disable the bot, don't keep paying for it while ignoring it.

Reflexively addressing. "BugBot said something, must fix." That treats BugBot as authoritative, which it isn't. The fixer is wrong sometimes. Read the comments; act on the real ones.

The middle path — treat BugBot like any other reviewer, address valid feedback, dismiss invalid feedback, learn over time — is what makes the tool useful.

The 80% Resolution Rate

The marketing number for BugBot 2.0 (the fixer release) is "80% issue resolution." What that means in practice:

- BugBot identifies an issue and opens a fix PR.
- For 80% of those issues, the fix PR merges as-proposed or with minor edits.
- For 20%, the fix PR closes without merging — the bot's proposed fix wasn't right, the issue wasn't real, or the team decided not to address it.

The 80% rate is high enough that BugBot meaningfully shifts work — most issues get fixed without human implementation. It's also not 100%, which means human review of fix PRs remains required. Treating BugBot fix PRs as auto-merge candidates is the failure mode.

A reasonable workflow:

1. BugBot opens a fix PR.
2. The PR's author (or a designated reviewer) reads the fix PR.
3. If the fix is right, merge.
4. If the fix is wrong, close with a note (which feeds back into BugBot's calibration).
5. If the fix is partially right, take what's useful and discard the rest.

This treats BugBot as a junior contributor whose work needs review. The 80% rate means the review is mostly approve-and-merge, which is fast — but the review still happens.

BugBot vs. Human Review

A common question: does BugBot replace human reviewers?

The honest answer is no, but it shifts the kind of review humans do. With BugBot handling the first pass for mechanical issues, human reviewers can focus on:

- **Design and architecture.** Is this the right approach? Does it fit the system?
- **Business logic.** Does the code do what it's supposed to do? Are the requirements correct?
- **Code clarity.** Is this readable to someone without the context?
- **Cross-cutting concerns.** Does this PR conflict with other work in flight? Does it set precedent we'll regret?

These are higher-value questions than "did you forget to await this Promise" — and they're exactly what humans do better than automated tools. Done well, BugBot raises the quality of human review by removing the mechanical burden.

Common BugBot Pitfalls

Three patterns to avoid:

Enabling BugBot on every repo without piloting. Different repos have different cultures and conventions. Pilot on one, learn the configuration, then expand.

Ignoring comment volume. If BugBot generates 20+ comments per PR, the configuration is too noisy. Tune severity thresholds, disable irrelevant categories, tighten project rules.

Treating BugBot's confidence as truth. BugBot writes its comments confidently. The confidence isn't calibrated to truth. Read the content, not the bot's tone.

From Bot to Builder

BugBot is one example of an automated workflow plugging into Cursor's ecosystem. Chapter 24 covers the local extensibility primitives — hooks, slash commands, and Skills — that let you tune the editor for your team's specific workflows, including the Skills pattern that bundles rules with templates and slash commands into a reusable unit.

Chapter 24: Hooks, Slash Commands, and Skills

Beyond rules and MCP, Cursor exposes three more extensibility primitives that shape how the editor behaves for your specific workflows: hooks (event-driven shell scripts), slash commands (named prompts you invoke), and Skills (the rules-plus-templates pattern that bundles rules, slash commands, and reference assets into a reusable unit). Together they're the layer where Cursor stops being a generic editor and becomes a tool tuned to your team.

This chapter covers each, where they earn their keep, and how teams share them.

Hooks

Hooks are shell scripts Cursor runs in response to specific events. The mental model is the same as git hooks: a directory of named scripts that fire at predictable points in the editor's lifecycle. Hooks live in `.cursor/hooks/` at your project root.

Common hook events:

- **Pre-prompt.** Before any AI invocation. Useful for adding context (current branch, current ticket ID, environment) to every prompt.
- **Post-edit.** After an AI invocation produces an applied change. Useful for triggering formatters, linters, or notifications.
- **Pre-commit.** Before a commit. Same as git's pre-commit but with awareness of which changes were AI-generated.
- **On-tool-call.** When an MCP tool is about to be invoked. Useful for additional approval logic or logging.
- **Session-start.** When a Cursor session opens on the project. Useful for environment setup or warning banners.

A hook is a small executable script (shell, Python, Node — anything you can run with a shebang). Cursor invokes it with relevant context as arguments and environment variables; the script does its work and returns exit code 0 (allow/continue) or non-zero (block/error).

A simple example — a pre-prompt hook that adds the current ticket number to every Agent prompt:

```
#!/bin/bash
# .cursor/hooks/pre-prompt
TICKET=$(git branch --show-current | grep -oE '^[A-Z]+-[0-9]+')
if [ -n "$TICKET" ]; then
  echo "Current ticket: $TICKET"
fi
```

Cursor reads the hook's stdout and prepends it to the model's context. Every Agent prompt now knows the ticket from your branch name. The hook is small, durable, and applies team-wide once you check it in.

Another example — a post-edit hook that runs the project formatter on edited files:

```
#!/bin/bash
# .cursor/hooks/post-edit
for file in "$@"; do
  case "$file" in
    *.ts|*.tsx) npx prettier --write "$file" ;;
    *.py) black "$file" ;;
    *.go) gofmt -w "$file" ;;
  esac
done
```

Now every AI edit gets formatted automatically. No more "the model wrote it but didn't run the formatter" friction.

When to use hooks:

For automation that must always happen. Formatters, linters, security scanners, telemetry. Things you don't want to rely on the model or the user to remember.

For team-wide enforcement of conventions. A pre-commit hook that blocks commits with debug `console.log` calls left in. A pre-prompt hook that loads project context every time. Once checked in, everyone gets the behavior automatically.

For integration with your team's tooling. A hook that posts a summary to Slack when a Background Agent finishes. A hook that updates Linear when a commit references an issue. The MCP-vs-hook split: MCP is for capabilities the model invokes; hooks are for behaviors that fire automatically.

When not to use hooks:

For things that should be optional. A hook is unconditional — it fires on every event of its type. If the behavior should be conditional or opt-in, it's a slash command, not a hook.

For long-running operations. Hooks block the event they're attached to. A 30-second hook on every pre-prompt makes the editor feel sluggish. Keep hooks fast or run their work asynchronously.

For one-off scripts. If the hook only matters for one engineer's workflow, don't check it into the team's shared `.cursor/hooks/`. Use a personal global hook instead.

Slash Commands

Slash commands are named prompts you invoke from chat. Type `/` in the chat input and Cursor shows a menu of available commands; pick one, optionally provide arguments, and the command's prompt template runs.

Slash commands live in `.cursor/commands/` as markdown files. Each file defines one command:

```
---
name: review-this-pr
description: Review the current PR carefully and post a summary
---

Read the current PR's diff and changes. Identify:
- Any bugs or potential issues
- Style or convention violations
- Missing test coverage
- Performance concerns
- Security risks

For each issue, explain the concern and suggest the fix. Format as a checklist
```

```
suitable for posting as a PR comment. End with an overall assessment:
"Looks good," "Needs minor changes," or "Needs major rework."
```

In chat, type `/review-this-pr`. Cursor expands the slash command into the prompt above and sends it. You skip the typing — and more importantly, the team's idea of "what does a thorough PR review look like" is encoded in the command, so every engineer running `/review-this-pr` produces the same kind of review.

Slash commands can take arguments:

```
---
name: write-test-for
description: Write a test for the named function
arguments:
  - name: function_name
    description: Name of the function to test
    required: true
---

Write a comprehensive test for the function `{function_name}` in this codebase.
The test should:
- Cover the happy path
- Cover edge cases (empty input, boundary conditions)
- Mock external dependencies appropriately
- Match the project's testing conventions
```

Invoke with `/write-test-for searchProducts` and the command runs with `function_name` substituted.

When slash commands earn their keep:

Repeated workflows. "Generate a CHANGELOG entry for these commits." "Refactor this for testability." "Audit this PR for security." If you find yourself typing the same prompt skeleton repeatedly, it's a slash command.

Team-shared workflows. When your team has a specific kind of work — release prep, post-incident review, deploy steps — a slash command captures the team's recipe. New engineers get the command for free; everyone runs the same template.

Multi-step prompts you don't want to retype. Long prompts with specific structure are tedious to remember and type. Slash commands serve as personal or team libraries.

When slash commands are wrong:

One-off prompts. Don't make a command for something you'll use once. Use the prompt directly.

Trivial prompts. "Add a docstring" is faster to type than `/add-docstring`. Slash commands are for prompts that benefit from templating.

Prompts that need substantial customization per use. If 80% of the slash command is parameters or context the user has to provide each time, the command isn't templating much. Just write the prompt.

Skills

Skills are the higher-level pattern — a directory that bundles rules, slash commands, and reference assets into a unit you install or share. Where rules are individual configuration files and slash commands are individual prompts, a Skill is a coordinated package.

A typical Skill directory:

```
.cursor/skills/release-prep/
├── SKILL.md              # Skill definition, description, when-to-use
├── rules/
│   └── release-conventions.mdc
├── commands/
│   ├── changelog.md
│   ├── version-bump.md
│   └── release-notes.md
├── templates/
│   ├── release-checklist.md
│   └── changelog-template.md
└── README.md             # Documentation
```

The `SKILL.md` is the entry point. It describes what the Skill does, when to use it, and what it includes:

```
---
name: Release Prep
```

```
description: Workflow for preparing a release — version bump, CHANGELOG, release
 notes
---

# Release Prep Skill

Use this Skill when preparing a release.

## What it includes

- Project rule: release conventions for this repo
- Slash commands:
  - `/changelog` — generate a CHANGELOG entry from recent commits
  - `/version-bump <type>` — bump version per semver
  - `/release-notes` — draft external-facing release notes
- Templates: standard release checklist and CHANGELOG format

## How to use

1. Run `/changelog` to update CHANGELOG.md.
2. Run `/version-bump minor` (or major/patch) to bump the version.
3. Run `/release-notes` to draft the announcement.
4. Follow `templates/release-checklist.md` for verification steps.
```

When the Skill installs, the rules apply to relevant files, the slash commands become available, and the templates are accessible to any prompt that needs them. Engineers see the Skill as a unit; they don't have to know which file does what.

Skills are particularly powerful for:

Onboarding. A new engineer joins the team. Instead of reading 12 different rule files and hoping they pick up the conventions, they install the team's Skills. Each Skill is self-documented; the engineer learns the workflow by running the commands.

Cross-team sharing. If your team has a particularly good release-prep workflow or a great PR review pattern, packaging it as a Skill lets you share with other teams or open-source it for the community.

Standardization. When the org has many teams, Skills become the unit of cross-team standardization. "All teams use the security-review Skill before merging anything that touches auth."

Personal libraries. Engineers build personal Skills for their daily workflows. A "morning routine" Skill (open issues, review PRs, etc.). A "Friday wrapup" Skill (CHANGELOG, status update, weekend prep).

Sharing Across Teams

The shared Skill pattern — where one team builds something and other teams consume it — has a few practical forms:

Org-wide repository of Skills. A `cursor-skills` repo with shared Skills. Teams pull from it as a git submodule or sync periodically. The maintainer team curates; consumer teams subscribe.

Per-team Skills shared via internal package. Some teams package Skills as installable npm or pip packages and version them. Heavier setup, but works well for organizations with strong package-management culture.

Public Skills via open-source. Teams that want to share publicly publish Skills as GitHub repos. The community version is growing — several public repositories of Cursor Skills exist as of mid-2026.

Skills are versioned. When a Skill ships an update, consumers update on their schedule. The convention is semver: breaking changes (rule rewrites, command renames) bump major; additive changes bump minor.

Walkthrough: A Custom Slash Command for Deploy

Concrete example. Your team's deploy workflow has six steps the engineer doing the deploy needs to run. You codify it as a slash command:

Step 1: Create `.cursor/commands/deploy.md`:

```
---
name: deploy
description: Walk through the team's deploy workflow
arguments:
  - name: env
    description: Environment to deploy to (staging, production)
    required: true
---

You are helping the developer deploy to {env}.
```

```
Step 1: Confirm the current branch is `main` (or a release branch for productio
n).
If not, stop and tell the developer to switch.

Step 2: Confirm there are no uncommitted changes. If there are, stop.

Step 3: Run `npm run build` and confirm it succeeds. Show the output.

Step 4: Run `npm run test:integration` and confirm all tests pass. Show the outp
ut.

Step 5: Run `./scripts/deploy.sh {env}` and watch the output. If the script fail
s,
stop and surface the error with diagnosis.

Step 6: After deploy succeeds, post a brief summary to the team's Slack channel
using
the slack MCP tool. Format: "Deployed [commit-sha] to {env}. [link to deploy lo
g]."

Walk through these steps one at a time. Confirm with the developer between steps.
After step 6, mark the deploy complete.
```

Step 2: Commit and push. The team gets the command on next pull.

Step 3: Use it. On deploy day, type `/deploy staging` in chat. Agent walks through the steps, runs the commands, asks for confirmation, posts to Slack at the end.

Step 4: Iterate. After a few deploys, the team notices step 4 is unreliable on slow CI. Update the command to add a retry with timeout. The next deploy gets the improvement automatically.

The slash command becomes the team's living deploy runbook. New engineers don't need a separate doc; they have the command.

Hooks vs. Slash Commands vs. Skills

A summary of when each is the right primitive:

Hooks — fire automatically on events. Use for unconditional behaviors that must always happen (formatting after edits, security checks before commits, telemetry).

Slash commands — invoked deliberately by the user. Use for repeatable templated prompts that you want to run easily but don't want firing automatically.

Skills — bundle rules + commands + templates. Use when you have a coherent workflow that benefits from being packaged together.

The progression is natural: a project starts with rules, accumulates a few slash commands, grows hooks for automation, and eventually a particularly important workflow gets packaged as a Skill. Don't over-engineer the structure upfront; let it grow as the team's needs emerge.

Maintenance and Hygiene

A few practices that keep this layer from becoming legacy debt:

Document each artifact. Rules, commands, hooks, Skills — each should have a clear description and "when to use." Without docs, the team forgets why an artifact exists, and inertia keeps it around long after it's stopped being useful.

Periodically review. Quarterly cleanup of `.cursor/` is a good cadence. Anything unused in 90 days, evaluate. If still useful, document. If not, remove.

Version Skills explicitly. When you change a Skill in a way that affects consumers, bump the version. Document the change. The cost is small; the benefit when something breaks downstream is large.

Don't put secrets in any of these files. Hooks can read environment variables; commands can reference them; Skills should reference them. Don't check secrets into the `.cursor/` directory — it's shared with the team and possibly public.

Treat the directory like code. PR review for changes. Tests where applicable. Don't change rules without telling the team. The `.cursor/` directory is part of the codebase, not a personal scratchpad.

Rules, MCP, hooks, slash commands, Skills — that covers the major extensibility surface. Chapter 25 closes Part VI with the runtime layer: how Cursor's terminal differs from external terminals, how to give Agent the right scope for command execution, and AI-assisted debugging.

Chapter 25: Working with the Terminal and the Debugger

Cursor inherits VS Code's terminal and debugger, then adds AI on top. The terminal is where you run things, where Agent runs things on your behalf, and where the most consequential approval decisions happen. The debugger is where AI assistance shifts from "helps me write code" to "helps me understand what this code is doing." This chapter covers both.

The Terminal

Cursor's integrated terminal is the same terminal you've used in VS Code. A fully functional shell embedded in the editor — bash, zsh, fish, PowerShell, or whatever you've configured. Cmd+` (Ctrl+`) opens it; the standard VS Code shortcuts and commands all work.

The Cursor-specific affordances:

Inline AI on terminal output. Right-click any line of terminal output and select "Ask AI about this." Cursor opens a chat with the line attached as context. Useful for "why did this command fail" or "what does this error mean."

Terminal commands as chat context. When you reference recent terminal output in chat, Cursor reads the terminal buffer as context. "The last command failed; help me debug it" works without manually copy-pasting output.

Agent-driven commands. When Agent mode runs commands, they execute in a Cursor-managed terminal session you can watch in real time. The output joins Agent's working context; you also see it in the terminal panel.

Per-session command approvals. The first time Agent wants to run a command in a session, Cursor prompts for approval. Grant for the session, grant once, or deny.

Cursor's Terminal vs. an External Terminal

A common question for new users: should I use Cursor's terminal or my external terminal?

Cursor's terminal:

- Integrated with the editor; AI sees its output.
- Agent can drive it.
- Recent commands available as context for chat.
- Same shell as your normal one (uses your shell config).

External terminal:

- Doesn't share state with Cursor.
- Agent can't drive it.
- Better for workflows where you want isolation (production-touching commands, things you don't want AI to see).
- Familiar muscle memory if you've been using one for years.

The book's recommendation: use Cursor's terminal for development work that benefits from AI integration, and an external terminal for things you don't want in Cursor's context (production deploys, sensitive credentials, anything you want fully manual). The split is a privacy and safety practice as much as a tooling preference.

For Agent and Background Agent runs specifically, the AI runs commands in Cursor-managed terminals — no choice there. The choice is what you do yourself.

Agent's Command Execution

When Agent mode wants to run a command, the flow:

1. Agent identifies a command to run as part of its loop.
2. Cursor checks whether the command is approved for the session.
3. If not, Cursor prompts you for approval.
4. On approval, the command runs in a Cursor-managed terminal.
5. Output streams back to Agent's context and to the terminal panel.

6. Agent continues based on the output.

The approval prompt is your safety net. It shows the command Agent wants to run, lets you approve once or for the rest of the session, and gives you the option to deny. The granularity matters:

Approve once. This specific command runs. The next command Agent wants to run prompts again.

Approve for session. This command and similar ones run without prompting until the session ends.

Approve always (for this command). Permanently approve this exact command for this project. Cursor remembers and won't prompt again.

Deny. This command doesn't run. Agent has to find another way (or stop).

The book's recommendation: don't auto-approve everything. The default per-command prompt is friction, but it's the kind of friction that prevents disasters. Approve the obvious-safe things (running tests, formatting, building). Be more careful with anything that touches state — database operations, deploys, file deletions, anything that reaches outside your repo.

Safety Knobs

Cursor exposes several knobs for controlling what Agent can do in the terminal:

Allowlist of safe commands. Settings → Agent → Allowed Commands. Pre-approve specific commands (`npm test`, `npm run typecheck`, `git status`) so Agent doesn't have to prompt for them. Speeds up Agent loops without compromising safety on actually-dangerous commands.

Blocklist of dangerous commands. Settings → Agent → Blocked Commands. Patterns like `rm -rf`, `sudo`, `git push --force` can be blocked entirely; Agent can't run them even if you approve. Prevents the catastrophic-error category at the harness level.

Per-project command policies. Repos can have their own allow/block lists in `.cursor/agent-policy.json`. The team-shared file means everyone's Agent has the same safety boundaries.

Confirmation requirements. Some commands always require a fresh approval per call regardless of session-level approvals. Useful for "this command is conditionally safe" cases — `git push` to your dev branch is fine; `git push` to main isn't.

Background Agent's separate policy. Background Agent runs in a sandbox with a more permissive command set by default (it's isolated from your local filesystem). Policies there are configured separately.

The Debugger

Cursor inherits VS Code's debugger — full-featured, with breakpoints, watch expressions, call stacks, variable inspection, conditional breakpoints, and the rest. The Cursor-specific affordances are the AI-assisted ones layered on top:

AI-assisted breakpoint setting. Right-click in the gutter and "Suggest Breakpoint" asks the model where to place breakpoints based on your debugging goal. Useful when you're new to a codebase and not sure where to start.

Variable explanation. Hover over a variable in the debug view and "Explain this value" asks the model what the value means. Useful for opaque values — base64 strings, structured data, IDs that map to something.

Call stack analysis. Select a frame and "Explain this stack" returns a model-written explanation of how execution got here. Useful for unfamiliar libraries or deep stacks.

Failure hypothesis generation. When a test fails or an exception throws, "Why might this fail?" returns the model's hypotheses. Often saves time on the "could it be X? could it be Y?" enumeration phase of debugging.

These aren't replacements for understanding what your code does. They're shortcuts for the parts of debugging that benefit from a confident first guess. The model's hypotheses are starting points; verifying them remains your work.

The AI-Assisted Debugging Workflow

A pattern that combines the debugger and AI primitives effectively:

Step 1: Reproduce. Get the failure to happen reliably. AI doesn't help here; this is environment and reproduction work.

Step 2: Set initial breakpoints. Either manually or via "Suggest Breakpoint." Place them where the failure becomes visible — usually one breakpoint upstream of the error and one at the error itself.

Step 3: Run to breakpoint. Step through. Inspect variables. Build a picture of the state at each breakpoint.

Step 4: Hypothesize. When you have enough state to reason about, switch to chat (Ask mode) with the variable values pasted in. "Given these values at this breakpoint, what could cause the error at the next breakpoint?" The model's hypotheses are usually a productive starting point.

Step 5: Test the hypothesis. Either change code to validate (if cheap) or set more breakpoints to inspect specific values.

Step 6: Fix. Once the cause is identified, write the fix — Cmd+K for inline edit, Manual for coordinated fixes, or type it yourself if it's small.

Step 7: Re-run. Confirm the fix. Add a regression test if one wasn't there.

The same debugging workflow you'd do without AI, but with the model accelerating the hypothesis phase. For novel bugs in unfamiliar code, the speedup is significant.

Walkthrough: Debugging a Flaky Test with Agent

Concrete example — using Agent for autonomous debugging.

Scenario: A test fails intermittently in CI. Locally it passes most of the time but fails one in twenty. The team has been ignoring it for two weeks.

Step 1: Open the test file in Cursor.

Step 2: Switch to Agent mode. Send:

> "The test `tests/orders.test.ts → orders.placeOrder.handles concurrent orders` is flaky. It passes most of the time but fails one in 20 runs. Reproduce the failure, diagnose it, and fix it. Run the test 50 times after fixing to confirm it's stable. Don't change the test's intent —

only the implementation, the test setup, or the test itself if the test was wrong."

Step 3: Watch. Agent runs the test in a loop. The first 15 runs pass. Run 16 fails — Agent captures the failure output, reads the test, reads the implementation it tests, and forms a hypothesis: a race condition in the order ID generation.

Step 4: Agent proposes a fix. It identifies that two concurrent calls can generate the same order ID — timestamp-based ID without a uniqueness suffix. It proposes adding a sequence counter.

Step 5: Agent implements. The diff is reviewable in real time as Agent works. The change is small — a few lines in the ID generator, a few lines in the test setup to verify uniqueness across concurrent calls.

Step 6: Agent verifies. Runs the test 50 times. All pass. Reports back: "Diagnosed the flake as a race condition in the order ID generator. Fixed by adding a sequence counter. 50 consecutive passes after the fix."

Step 7: You review. Read the diff. The fix looks reasonable. The test additions are sensible. Run the tests yourself to double-check. Commit and push.

The whole exercise might take Agent 15–30 minutes and you 5–10 minutes of supervision. The same task by hand could take an hour or more — much of it the boring work of running the test in a loop until it fails. Agent's read-act-observe loop is built for exactly this kind of patient repetition.

The discipline that makes it work: a clear scope (reproduce, diagnose, fix, verify), explicit success criteria (50 consecutive passes), and a constraint (don't change the test's intent). Without these, Agent might fix the flake by deleting the test or weakening its assertions — technically a "fix," but not the one you wanted.

The `.cursor/environment.json` File (Revisited)

We touched `.cursor/environment.json` in Chapter 16 in the Background Agent context. The file matters here too — it's where you tell Cursor (and Background Agent specifically) how to set up the environment for command execution.

A complete example for a TypeScript project:

```json
{
  "snapshot": "node-20-postgres",
  "install": "npm ci && cp .env.example .env",
  "start": "docker compose up -d postgres redis && npm run dev",
  "terminals": [
    {
      "name": "tests",
      "command": "npm test -- --watch"
    }
  ],
  "env": {
    "NODE_ENV": "development"
  }
}
```

The `snapshot` field references a cached disk image so the sandbox boots quickly. The `install` script runs once when a fresh sandbox is provisioned to install dependencies. The `start` command launches whatever services and processes the project needs. `terminals` defines background processes that run alongside the agent — a watch-mode test runner here. `env` sets environment variables inside the sandbox.

For projects without an `environment.json`, Cursor infers from common config files (`package.json`, `pyproject.toml`, `Cargo.toml`, `go.mod`). The explicit file is more reliable, especially for projects with non-standard setup.

Common Pitfalls

A few patterns to avoid:

Auto-approving everything Agent asks for. The friction of the per-command prompt is the safety net. Approving without reading is how the catastrophic-error category sneaks in.

Letting Agent run destructive commands without thought. `rm -rf`, `git push --force`, `DROP TABLE`. These deserve more attention than `npm test`. Block them at the policy level if you're worried.

Running commands manually that Agent should be running. If the workflow is "Agent makes changes, you run tests manually," you've split the loop. Either Agent runs tests too, or stay in Manual mode.

Ignoring command output. Agent's "tests pass" claim is text it generated. The actual test suite's exit code is the truth. Verify via the terminal, not via Agent's report.

Treating the debugger as a last resort. AI-assisted debugging beats print statements for substantive bugs. Reach for the debugger earlier than habit suggests.

Part VII: Cursor for Your World

"The right setup is the one that fits the work you actually do, not the work you imagine doing."

— A working principle

Chapter 26: Cursor for Solo Developers

The solo developer running Cursor is the most common reader of this book. You're working alone — a freelancer, a contractor, a startup founder, an indie hacker, a side-project builder, or an employed engineer paying for a personal Pro subscription on top of whatever your job provides. The setup that makes Cursor leverage for you differs from what works for a 50-person engineering team, and this chapter is about getting the solo setup right.

The thread running through the chapter: solo Cursor is about discipline. No team enforces conventions, no shared rules library someone else maintains, no admin watching credit usage. You're the convention-maker, the rules-writer, and the budget-watcher. Done well, it's leverage. Done poorly, it's expensive frustration.

The Solo Pro Subscription

Most solo developers should be on Pro at $20/month. The cheaper tier (Hobby) is functional but limiting; the more expensive tiers (Pro+, Ultra) are usually overkill until you've established a workflow that justifies them.

The Pro plan's $20 of monthly credits is genuinely enough for a productive month, *if* your workflow has discipline around model selection. The "$20 in three days" trap is the most common solo failure mode, and earlier chapters laid out the discipline that prevents it. The short version:

- DeepSeek V4 Pro by default for chat and Manual.
- Opus 4.7 reserved for cases that genuinely need it — gnarly debugging, architecture decisions, high-stakes refactors.
- Agent used selectively, not as the default mode for every task.
- A hard budget cap set at $30/month for the first few months while you calibrate.

Most solo users who follow this discipline find Pro comfortably sufficient. Those who upgrade to Pro+ usually do so because their volume genuinely outgrew Pro, not because they wasted credits.

The Solo Workflow

A typical solo developer day with Cursor:

Morning (light usage): Tab autocomplete handles small edits. Cmd+K for the deliberate medium ones. Maybe one or two Ask sessions to scope a task. Credit cost: minimal.

Implementation block (heavier usage): A focused task — a feature, a refactor, a bug fix. Manual mode handles the bulk; Agent for the verification-heavy parts. Credit cost: $1–3 for the block.

Investigation or design (variable): Hard problems where you spend time in chat on the harder Opus model. Higher per-round cost, lower volume. Total: $1–5 if needed.

Wrap (light usage): Code review on your own work, drafting a PR description, updating documentation. Mostly Ask mode and Cmd+K. Total: minimal.

A productive day typically lands at $2–5 in credit consumption. A $20 monthly budget covers 4–10 productive days at that rate. For most solo developers, that matches the actual cadence of focused implementation work — there are always days that are mostly meetings, reviews, planning, or other non-implementation work.

If daily consumption sits consistently above $5, the question is whether the work justifies it (in which case Pro+ at $60/month is the right tier) or whether the workflow has drifted (in which case better picker discipline brings the bill down).

Avoiding the Credit-Burn Trap

The pattern that catches new Pro subscribers:

1. Sign up. $20 of credits, "Auto" model, Agent mode for everything because it's exciting.
2. Day one feels like magic. Day three feels like magic. Days five through seven feel like magic.
3. Day eight: out of credits. Confused — "I just paid for a month."
4. Either upgrade in frustration or write off Cursor as expensive.

The fix is the discipline from earlier chapters, applied consistently. The specific habits for solo developers:

Manually pick the model. Don't leave it on Auto for the first month. Force yourself to think about which model fits each task. The friction is small and it builds the picker intuition that saves credits long-term.

Default to Manual mode for daily work. Use Agent only when the task genuinely benefits from autonomous execution. Most tasks don't.

Watch the credit dashboard. Once a day, check the usage panel. Notice which model and which mode consumed the most. If something stands out, ask whether it was the right tool for the work.

Set the hard budget cap. $30/month for the first few months. The cap protects you from runaway sessions; it also gives you a clear signal when you're approaching it ("you're at 80% of cap on day 18 — adjust").

Use Tab and Cmd+K aggressively for small work. Tab is essentially free. Cmd+K on cheaper models is nearly free. The bulk of small edits should never touch the credit pool.

A solo developer who internalizes this discipline within two weeks rarely struggles with credits again. The discipline isn't restrictive — it's calibration. You're learning what each feature costs and choosing accordingly.

Personal `.cursor/rules/` Libraries

Solo developers don't have a team's shared rules library. They have their own. Building one over time is high-leverage:

Per-project rules in each project's `.cursor/rules/` directory. These travel with the project and apply only there.

Personal global rules at the user level (Cursor settings → Global Rules). These apply across all your projects. Right for personal style preferences — naming conventions you use everywhere, comment styles you follow universally, libraries you reach for by default.

Rule templates in a personal `.cursor-templates/` repo (or wherever you keep templates). When you start a new project, copy in the relevant rule files as a starting point. Build the template library over the projects you work on; each new project inherits the accumulated wisdom of the old ones.

A typical solo developer's rule library, built over a year of use:

```
~/.cursor-templates/
├── typescript-react.mdc
├── python-fastapi.mdc
├── go-grpc.mdc
├── general-style.mdc
├── testing-conventions.mdc
├── git-conventions.mdc
└── README.md
```

Each rule captures something the developer reaches for repeatedly. New project? Copy the relevant ones. Update the template when you discover a new convention worth standardizing.

The book's recommendation: when you find yourself fighting the model for the third time on the same convention, write a rule. Five minutes writing a rule pays back in days of avoided friction.

Side Projects vs. Paid Client Work

Many solo developers split their Cursor time between paid client work (where leverage compounds into revenue) and side projects (where they're learning or building for fun). The two need slightly different discipline.

For paid client work:

- The stakes justify higher model use. Opus on the hard parts is fine; time savings turn into billable hours saved.
- BYOK against a Claude or OpenAI key can make sense at scale. If a single client engagement consumes substantial credits, BYOK saves real money.
- Privacy mode should always be on. Your client's IP isn't yours to leak through default data handling.
- Project rules are part of the deliverable. If you're handing off the codebase, the rules go with it.

For side projects:

- DeepSeek for almost everything. Opus rarely. The marginal quality of premium models doesn't matter on hobby code.
- Privacy mode is optional. If the project is open-source or doesn't matter, skip the marginal protection.
- Shorter sessions. Side project time is finite; don't burn credits and time on long Agent runs going nowhere.
- Use side projects to test new Cursor features. Try Multitask, try BugBot on your personal repo, try a new MCP server. The side project is the safe laboratory.

The split matters because the right discipline depends on whether you're optimizing for output (client work, where every productive hour matters) or for learning (side projects, where the journey is the point).

When You're the Only Reviewer

Solo developers don't have a teammate reviewing their PRs. The "review every line" discipline is more important, not less, because the model's mistakes don't get caught by a second human. Some practices:

Use BugBot on personal repos. BugBot Pro with usage-based billing on your repos. The cost is small for low-volume PR queues, and BugBot catches things you'd miss after staring at the code for a long session.

Write tests before you trust Agent output. Agent's confidence is calibrated to its training, not to your project's reality. Your test suite is how you verify. If the test suite isn't strong enough, Agent's output isn't trustworthy regardless of its "all tests passing" claim.

Cool-down before merging. Don't merge an Agent-produced PR five minutes after it lands. Give yourself an hour, ideally a night, to come back with fresh eyes. The bugs the model misses often look obvious on a second pass.

Periodic codebase audits. Once a quarter, ask Ask mode to review your codebase for inconsistencies — places where the rules aren't followed, dead code, deprecated patterns. The audit takes an hour and catches the drift solo work accumulates.

A Solo Weekend Project Walkthrough

Concrete example. Saturday morning, you decide to build a simple URL shortener as a weekend project. The tech stack is TypeScript + Hono + SQLite. You've never used Hono before.

Saturday morning:

- Open a fresh project. Set up a minimal Hono boilerplate.
- Set Privacy mode on (habit, not strictly necessary for hobby work).
- Default model: DeepSeek V4 Pro.
- Hard budget cap for the project: $5.

Hour 1: Scoping.

Open chat in Ask mode. Ask:

> "I'm building a URL shortener with TypeScript and Hono. The features I want: anonymous URL submission, generated short codes, redirect from short to long URL, basic analytics on click counts. SQLite for storage. Walk me through the architecture and the components I'll need."

Read the answer. Push back on anything weird. End with a clear plan.

Hour 2: Scaffold.

Switch to Agent mode. Send:

> "Scaffold a Hono application with the structure we just discussed. Set up SQLite via a lightweight library appropriate for this use case. Don't implement business logic yet — just the structure with placeholder handlers. Run the dev server to confirm it starts."

Watch Agent set up the project. Review the diff. The structure looks reasonable.

Hour 3–4: Implement.

Work through the feature list in Manual mode: URL submission, short-code generation, redirect, click counting. Each is a focused multi-file change reviewed and accepted.

Hour 5: Polish.

Switch back to Ask mode. "Look at the project. What's missing for a basic launch? Suggest improvements that would matter." Use the suggestions to drive a final round of Manual edits — error handling, edge cases, basic styling.

Hour 6: Deploy.

Open the terminal. Deploy to Fly.io or wherever. Test in production.

Total credit consumption: roughly $2–3 for the day. The project ships. You learned Hono. The Cursor leverage compounded — you'd have spent twice as long doing the same project without it, and learned less Hono because you'd have spent more time fighting boilerplate.

Personal MCP Configuration

A solo developer's MCP setup is leaner than a team's:

Global servers for tools you use across projects. A personal note-taking MCP (Notion, Obsidian, etc.). A web fetch server. A sequential-thinking server.

Per-project servers scoped to where they apply. Postgres for projects with a database. GitHub for projects with PRs. Linear if you track personal tasks there.

Total tool count typically lands at 10–20 across all projects. Well below the 40-tool ceiling and small enough to remember what each does.

The book's recommendation for solo developers: don't install MCP servers preemptively. Add them when you've felt the lack of them twice. "Just in case" servers add tool-list noise without daily value.

Privacy and Solo Work

Solo developer privacy considerations:

Client work. Always Privacy mode on. Your client's code isn't yours to share through default data handling.

Open-source work. Privacy mode is optional. The code is public; the marginal protection isn't load-bearing. Some prefer to leave it on as habit.

Personal projects. Optional. Same reasoning as open-source.

Anything containing secrets, credentials, or PII. Privacy mode plus aggressive `.cursorignore` for the sensitive files. Don't index a secrets file even if it's in your private repo.

Self-hosted/sensitive infra work. Consider whether Cursor is the right tool at all. If you're working on infrastructure where the code itself is sensitive (security tooling, financial systems, regulated work), the calculus shifts. Read Chapter 31 carefully before committing.

When Solo Becomes Team

A common transition: you start solo, the project grows, you hire your first engineer. The Cursor setup that worked for solo doesn't automatically scale to a team.

The transition checklist:

- **Move per-project rules from local-only to checked-in.** They go in `.cursor/rules/` in the repo. Both engineers benefit.
- **Audit personal rules for what should be project rules.** If your global rules contain conventions that apply to this project, copy them in.
- **Switch to Teams plan.** $40/user/month, pooled credits, shared admin. Worth it from the second engineer.
- **Add MCP configurations to `.cursor/mcp.json`.** Both engineers get the same tooling automatically.
- **Document the workflow.** A `CONTRIBUTING.md` or similar covering how the team uses Cursor — which mode for what, what the rules expect, what BugBot is configured to do.
- **Pilot BugBot.** Two engineers reviewing each other's PRs is the right moment for automated first-pass review.

The transition is mostly mechanical. The harder part is shifting the mental model from "my workflow" to "our workflow" — letting your conventions become the team's, which sometimes means compromise. Chapter 27 picks up there: shared rules, centralized billing, RBAC and SSO, onboarding, and the playbook for a 12-person team running Cursor effectively.

Chapter 27: Cursor for Teams

The Teams plan exists for a reason. A team of five engineers each on individual Pro plans misses most of the leverage Cursor offers a team — shared rules, centralized billing, pooled credits, unified MCP, role-based access, single sign-on. The $40/user/month Teams subscription pays for itself the first time a rule change ships to the whole team simultaneously instead of being explained five times.

This chapter is about getting the team setup right: which features matter, how to onboard new engineers without a long ramp, and the playbook for a team running Cursor as a workflow standard rather than as individual assistance.

Why Teams Beats Pro

The Teams plan changes Cursor from "individual editor with AI" to "team workflow with AI." The features that justify the upgrade:

Shared `.cursor/rules/`. Already supported on individual plans (rules are per-project, in the repo). Teams adds central administration of rule libraries — the admin can push rule updates to all repos, curate a starter library new repos inherit, and audit which rules each repo uses.

Pooled credits. Instead of per-engineer $20 caps, Teams pools credits across the org. One engineer's heavy refactor day doesn't cap them out while another's quiet meeting day leaves credits unused. The pool smooths the variance.

Centralized billing. One invoice, one payment method, expense reporting that finance can actually deal with. Individual Pro subscriptions on personal cards become an expense reimbursement nightmare past three or four engineers.

Usage analytics. The admin dashboard shows per-engineer usage, model preferences, credit consumption over time. Useful for spotting patterns — the engineer burning credits on Opus when DeepSeek would do, the engineer using only Tab when more advanced features could help.

Shared MCP configurations. Admin-installed MCP servers are available to everyone. Adding the team's Linear integration once means every engineer has it; you don't run setup five times.

SAML/OIDC SSO. Sign in via the team's identity provider (Okta, Azure AD, Google Workspace). Provisioning and deprovisioning follow your normal employee lifecycle.

Role-based access controls. Admins, members, and limited guest access patterns. Useful for contractor scenarios and for limiting who can change team-wide settings.

Privacy mode enforced org-wide. If the team's posture is "always Privacy mode," the admin enforces it. No engineer accidentally turning it off and using the editor on proprietary code without protection.

For a 3-engineer team, these features are nice-to-have. For a 5–10 engineer team, they're the difference between Cursor as a workflow standard and Cursor as a chaotic per-engineer choice. The book's recommendation: move to Teams as soon as you have 3+ engineers using Cursor seriously.

Shared Rules and Conventions

The single highest-leverage Teams feature is the shared rules library. The pattern that works:

Each repo owns its rules in `.cursor/rules/`. The repo's conventions are versioned with the code, reviewed in PRs, and travel with the codebase. Same as the solo case, now applied team-wide — every engineer who clones the repo gets the same rules.

The team has a shared "starter rules" repo. A separate repo containing rule templates the team uses across projects. New project? Pull the relevant starter rules as a baseline. The starter library captures cross-project conventions: how tests are named, how errors are handled, how commits are formatted, how code is documented.

Rule changes go through PR review. Treat rules like any other code. If an engineer wants to change a convention, they PR the rule change. The team reviews. If approved, it ships and everyone gets the update.

Quarterly rule audits. Every three months, sweep through the rule files. Anything obsolete? Anything that's drifted from how the team actually works? Update or remove. Maintenance is light but matters; rules that don't match reality erode trust in the rules system.

A team's rule library is one of the most valuable artifacts the team produces. New engineers can read the rules and learn the team's conventions in an afternoon. Senior engineers' tacit knowledge becomes explicit. The model's outputs become consistent across the team without requiring everyone to write everything by hand.

Centralized Billing and Usage Analytics

The Teams admin dashboard shows aggregate usage and per-engineer breakdowns. Some patterns to watch for:

Outliers in either direction. One engineer consuming 5x the team's average warrants a conversation — either there's a legitimate heavy-refactor reason or there's a workflow issue (always-Opus default, Agent for everything) the engineer doesn't see. One engineer consuming 1/10th the team's average might be a quiet user, or might indicate they're not using Cursor's higher-leverage features (Manual, Agent, rules) at all.

Per-model breakdowns. If 80% of the team's credits go to Opus, the picker discipline is missing. A team-wide nudge toward DeepSeek for routine work can drop the bill 50%+ without quality loss.

Per-mode breakdowns. Heavy Agent use is fine if the work justifies it; heavy Agent use on tasks that should have been Manual is wasted credits. Compare per-engineer breakdowns and identify outliers.

Per-repo usage. Some repos consume Cursor more heavily than others. Often the heavy ones are the ones where the team is actively building; sometimes they're the ones where the workflow has drifted. Diagnose case by case.

The point of analytics is calibration, not surveillance. The admin reading the dashboard should be looking for systemic improvements, not micromanaging individuals. Over-watching engineers' Cursor usage is a culture mistake.

Onboarding New Engineers

A new engineer joins the team. The Cursor onboarding should take an hour or two, not a week. The checklist:

Step 1: Account setup. Add the engineer to the Teams plan via the admin dashboard or via SSO/SCIM if configured. They get a sign-in link.

Step 2: Install Cursor. Cursor.com → download → install. If they were already on VS Code, the import flow carries their settings, themes, extensions, and keybindings over in one click.

Step 3: Privacy mode and basic config. Walk them through the first-launch settings. For most teams, Privacy mode is enforced org-wide, so this is a no-op. Confirm they're signed in to the team's GitHub organization.

Step 4: Clone the active project. Standard git clone. The repo's `.cursor/` directory provides project rules, ignore configuration, MCP setup, hooks, and slash commands automatically. The engineer doesn't have to set any of this up.

Step 5: Walk through the team's CONTRIBUTING.md (or equivalent). A short doc covering: - Which mode for which kind of work (Ask/Manual/Agent) - What the team's rules expect (or a link to the rules directory) - What BugBot is configured to do - Which MCP servers are in use and what they do - How the team handles credit budget (per-engineer limits, escalation paths)

Step 6: Pair with an existing engineer. Half a day of pair programming where the experienced engineer walks through their daily Cursor use. The new engineer sees the rhythm — when to switch modes, when to reach for Agent, when not to. Tacit-knowledge transfer no doc replaces.

Step 7: First task. A small, well-scoped task to ramp on. The new engineer uses Cursor as the team uses Cursor; the first PR is their calibration moment.

The whole onboarding takes a day, sometimes less. Compare to the alternative — onboarding to a custom team toolchain that takes a week — and the leverage is obvious.

Role-Based Access

Teams plans include three role tiers (specifics evolve, but the rough shape is consistent):

Admins. Full control over plan settings, billing, team membership, shared rules, MCP configurations, audit settings.

Members. Standard engineer access. Use the editor fully, install personal MCP servers, configure personal settings. Can't change team-wide configuration.

Limited/guests. Reduced access patterns for contractors or temporary collaborators. Often constrained to specific repos.

The book's recommendation:

- **One or two admins per team.** Engineering manager and a senior engineer is the typical pattern. More than two creates coordination overhead; a single admin creates a single point of failure.
- **Most engineers are members.** No reason to over-restrict.
- **Contractors get limited access.** Scoped to the repos they need; off the global rules library; often without Background Agent access (Background Agent's broader authority is contractor-inappropriate).

Audit role assignments quarterly. Departed contractors should be off the plan within a day of departure; new admin assignments should be deliberate, not accidental.

SAML/OIDC SSO

For teams with an identity provider (Okta, Azure AD, Google Workspace), SSO setup is straightforward:

Step 1. Admin enables SSO in Cursor's team settings. Cursor provides the SAML metadata or OIDC client details.

Step 2. The team's IdP admin creates a Cursor application using the metadata. Configures attribute mapping if needed.

Step 3. The IdP admin assigns the application to the right groups (engineering, contractors, etc.).

Step 4. Engineers sign in to Cursor via the IdP. New engineers only need to be in the right IdP group; provisioning is automatic.

The SSO setup pays back immediately on offboarding. When an engineer leaves, removing them from the IdP group also removes their Cursor access. No manual seat removal, no risk of forgotten access. For organizations with churn or lots of contractors, SSO is the offboarding mechanism that actually works.

MCP and Hooks at the Team Level

Teams should standardize MCP configurations and hooks. The pattern:

MCP in `.cursor/mcp.json`. Per-project, checked in. The repo's MCP setup is the team's MCP setup for that project.

Global MCP for personal use. Each engineer's personal global MCP (their note-taking, their personal task tracker) stays per-engineer.

Hooks in `.cursor/hooks/`. Per-project, checked in. Team-wide automation lives here.

Slash commands in `.cursor/commands/`. Per-project. The team's runbooks become commands.

Skills in `.cursor/skills/`. Per-project. Coordinated workflows packaged as Skills.

The principle: per-project configuration in the repo, personal configuration at the engineer level. Don't blur the line. A hook that fires for everyone needs to be in the repo (so everyone has it); a personal preference doesn't.

A 12-Person Team's Playbook

A real team's setup, presented as a model:

Plan: Teams. 12 seats at $40/user/month = $480/month.

Rules: Per-repo `.cursor/rules/` plus a shared starter rules repo. The starter library has 15 rule files covering language standards, testing conventions, security baselines, and PR conventions. New repos pull the relevant rules in as part of the project-init script.

MCP: GitHub, Linear, Sentry, Postgres (read-only), Slack (read messages, post notifications). All catalog installs. Configured in `.cursor/mcp.json` per repo.

Hooks: A pre-prompt hook adding the current ticket from the branch name. A post-edit hook running the project formatter. A pre-commit hook running the linter and a quick test pass.

Slash commands: `/changelog`, `/release-notes`, `/deploy`, `/security-review`, `/onboarding-followup`. Each is the codified version of a runbook the team had documented elsewhere.

Skills: Two team Skills — "release-prep" and "incident-response." Each bundles rules, commands, and templates for the corresponding workflow.

BugBot: Enabled on the main monorepo. Configured to comment on high and medium severity, open fix PRs only on high severity. Project rules reference BugBot's expected behavior so the model and BugBot don't fight each other.

Privacy mode: Enforced org-wide.

Credit budget: Pool of $480/month at the Teams base allocation, plus $200/month in pay-as-you-go the engineering manager has authorized. Average consumption: $550/month, well within budget. Outlier engineer (heavy refactor week): peak of $80 in a week, smoothed by the pool.

Onboarding: A two-page CONTRIBUTING.md covering the Cursor workflow. Half-day pair-programming session for new engineers. A `/onboarding-followup` slash command the new engineer runs at end of week one to summarize what they've learned and identify gaps.

Maintenance: Quarterly rule audit. Monthly review of credit usage with the team. A Slack channel where engineers post Cursor tips and discoveries.

Outcome: The team ships at a noticeably faster cadence than before adopting Cursor seriously. Onboarding takes a day instead of a week. PR review burden has shifted toward architectural questions because BugBot catches the mechanical issues. The team treats Cursor as part of the development environment, like the test runner or the type checker — boring, reliable, expected.

The playbook isn't unique. Many teams have arrived at similar setups. The point: effective team Cursor use is mostly mechanical — standard pieces, configured deliberately, maintained regularly.

Common Team Mistakes

A few patterns to avoid:

Letting Cursor adoption stay individual. Each engineer on personal Pro, no shared rules, no shared MCP. The team has Cursor; the team doesn't have a Cursor workflow. Fix: move to Teams plan, build the shared configuration.

Ignoring credit usage. Pool runs out mid-month, engineering complains, manager upgrades reactively. Fix: monitor monthly, set alerts at 75% and 90%, build the discipline before the cap is hit.

Over-engineered onboarding. A 30-page Cursor onboarding doc nobody reads. Fix: a two-page CONTRIBUTING.md plus pair programming. The doc covers the structure; the pair covers the rhythm.

Rules drift. Rules written in a flurry six months ago, codebase has moved on, rules don't match reality, engineers stop trusting them. Fix: quarterly audit, update or remove rules that no longer apply.

BugBot enabled without configuration. Default settings produce too many comments, engineers learn to ignore the bot, the bot's value evaporates. Fix: tune BugBot's severity thresholds and category filters; make the comments worth reading.

Teams works at the dozens-of-engineers scale. Above that, you're in Enterprise territory — SCIM, audit logs, custom contracts, security review, compliance posture. Chapter 28 covers when you need it, what's involved in a security review, and how to position Cursor for organizations with formal compliance requirements.

Chapter 28: Cursor for Enterprise

The Enterprise tier exists for organizations that need more than the Teams plan offers — formal security review, compliance attestations, SCIM provisioning, audit logs, custom contractual terms, and the operational maturity to manage AI tooling at org-wide scale. If you're in an organization with a security team, a procurement process, and a vendor approval workflow, Enterprise is the path.

This chapter covers what Enterprise actually includes, what a security review of Cursor looks like, and the operational practices that make Cursor work at org-wide scale.

When You Need Enterprise

Enterprise is the right fit when:

You have a formal security review process. Vendor approval requires SOC 2 attestations, DPAs, sub-processor lists, security questionnaires, and similar artifacts. Teams plans don't include the contractual machinery; Enterprise does.

Your IdP and lifecycle management are SCIM-driven. SCIM (System for Cross-domain Identity Management) is the protocol your IdP uses to automate user provisioning and deprovisioning. Teams supports SAML/OIDC SSO; Enterprise adds SCIM, the bar for org-wide identity hygiene.

You require audit logs. Comprehensive logging of AI invocations, repo access, configuration changes — the audit trail compliance frameworks demand. Teams has limited logging; Enterprise has full logs with retention policies.

You have repo-level access controls. Some repos shouldn't be indexable by Cursor at all (regulatory reasons, sensitivity, customer commitments). Enterprise lets admins configure repository, model, and MCP blocklists at the org level.

Your legal team needs custom data terms. Custom DPAs, geographic data residency, specific retention clauses — Enterprise contracts negotiate these. Teams uses standard terms.

Your compliance regime requires it. HIPAA workflows, FedRAMP-adjacent contexts, regulated financial workflows, defense-adjacent work. The Enterprise overlay is often what makes Cursor approvable in these contexts.

You're spending substantially. At dozens of seats, the cost difference between Teams and Enterprise becomes negligible relative to total spend, and Enterprise's flexibility in pricing and pooled usage often comes out cheaper.

If none apply, stay on Teams. Enterprise has overhead — sales cycle, legal review, operational requirements — that isn't worth it without a real driver.

The Security Review Process

Cursor's enterprise sales team has been through hundreds of security reviews and provides most of what your security team asks for proactively. Common artifacts:

SOC 2 Type 2 report. Cursor maintains SOC 2 Type 2 attestation. Your security team can review the report directly.

Sub-processor list. Which third parties Cursor's infrastructure uses (model providers, hosting infrastructure, telemetry vendors, etc.). Necessary for many security reviews because it determines who else is in your data flow.

Data flow diagram. What data leaves your environment, where it goes, what each hop does, what's retained where. Critical for understanding the privacy and risk picture.

Privacy mode behavior. Detailed documentation of what Privacy mode changes about retention. Often a key input to whether the security team approves use on sensitive workloads.

Encryption details. What's encrypted in transit, what's encrypted at rest, key management approach.

Incident response. How Cursor handles security incidents, notification timelines, customer communication during incidents.

Penetration test results. Recent pen test summaries (sometimes redacted). Demonstrates ongoing security validation.

Compliance certifications. SOC 2, possibly ISO 27001, possibly other regional certifications relevant to your industry.

Customer references in your industry. Other organizations under similar regulatory regimes using Cursor at scale. Useful for "are we the first?" reassurance.

Security review takes anywhere from two weeks (small organization, proactive review) to six months (large bureaucratic organization with sequential approval gates). Engage Cursor's enterprise team early; the materials they have are designed to address the common review questions efficiently.

SCIM Provisioning

SCIM is the protocol that lets your IdP create, update, and delete Cursor users automatically based on group membership in your IdP. The setup:

Step 1. Your IdP admin enables SCIM provisioning for the Cursor app. Generates SCIM credentials (typically a token).

Step 2. Cursor's enterprise admin enters the SCIM endpoint and credentials in Cursor's admin settings.

Step 3. The IdP starts pushing user changes to Cursor. New users in the engineering group get Cursor access automatically. Departed employees lose access immediately on IdP removal.

Step 4. Group-based role assignment maps IdP groups to Cursor roles. Members of `engineering-leads-okta` become Cursor admins; members of `engineering-okta` become members; members of `contractors-okta` get limited access.

The result: lifecycle management for Cursor matches lifecycle management for everything else. New hires don't sit waiting for Cursor access; departed employees don't retain it.

For organizations with frequent contractor turnover, SCIM is the only sustainable path. Manually managing seats per contractor scales poorly past a handful.

Audit Logs and Compliance Posture

Enterprise's audit logs capture:

Who accessed what. Per-engineer activity: which repos they invoked AI on, which MCP tools they called, which models they used.

When configuration changed. Rule additions, MCP installations, role assignments, plan modifications. Changes are logged with the actor and timestamp.

Sensitive operations. Privacy mode changes (if not enforced org-wide), credential resets, role escalations.

API access. Programmatic access via the Cursor API (where applicable) is logged separately.

Logs flow to your SIEM (Security Information and Event Management) system via:

- A pull-based API for SIEMs that periodically fetch
- A push-based webhook for SIEMs that prefer event-driven ingestion
- Direct integrations with major SIEM platforms (Splunk, Datadog, etc.)

Retention policies are configurable per contract. Most Enterprise contracts default to 12-month retention with longer available on request.

The logs serve two purposes: compliance attestation (your auditor wants to see logs for the review) and incident response (something went wrong and you need to know what happened). Set up SIEM ingestion early; you don't want to be scrambling to wire it up during an incident.

Repository, Model, and MCP Blocklists

Enterprise admins can constrain what Cursor can do at the org level:

Repository blocklists. Specific repos Cursor cannot index, regardless of user settings. Useful for repos with customer secrets, regulatory-bound code, or anything that contractually shouldn't touch AI tooling.

Model blocklists. Specific models Cursor cannot use. Useful when your organization has standardized on certain providers and doesn't want others called. "We don't use OpenAI; remove GPT from the picker."

MCP blocklists. Specific MCP servers Cursor cannot install or invoke. Useful for restricting Cursor's reach into systems you haven't approved. "Only catalog-curated MCP servers; no custom ones until reviewed."

Tool blocklists at the per-MCP level. Even for installed servers, specific tools can be blocked. "Postgres MCP is fine, but block the write operations."

These controls are policy-as-code. Enterprise admins configure them centrally; engineers can't override. The result: Cursor's authority is bounded by what the org has approved.

The book's recommendation for Enterprise admins: start with sensible defaults — block destructive operations on production systems, scope sensitive repos out of indexing — and add restrictions as engineering teams flag concerns. Over-restricting upfront produces a Cursor that engineers route around (using personal accounts, external tools); under-restricting is a security risk. Iterate toward the right level based on what your organization can support.

The AI-Code-Tracking API

Enterprise plans include an API for tracking AI-generated code in your repos. The capability:

- Each AI invocation in Cursor is associated with metadata about which engineer, which model, which mode produced the output.
- The metadata can be exported via the API.
- You can correlate it with git commits to identify which code was AI-assisted vs. human-only.

Use cases:

Compliance reporting. "What percentage of code shipped in Q2 was AI-assisted?" The API gives you the data.

Quality analysis. "Are AI-assisted changes more or less likely to have post-merge bugs?" Correlate the AI metadata with bug reports to learn about your tooling's effectiveness.

Audit responses. When a customer or auditor asks about AI use, you have authoritative data rather than estimates.

License compliance. For organizations concerned about copyright/license posture of AI-generated code, the tracking lets you identify AI-touched code if a question arises.

Most organizations don't need this initially; it becomes load-bearing as compliance requirements mature.

Pooled Usage and Invoice Billing

Enterprise contracts negotiate pooled usage at the org level rather than per-user. Specific patterns:

Pooled credit allocation. A monthly pool sized for the org's expected usage. Individual engineers don't have per-user caps; the org has a pool that covers everyone.

Usage-based pricing with volume discounts. Per-token rates negotiated below standard list prices for committed annual volume. Heavy users benefit from the volume.

Invoice billing. Net-30 or net-60 invoices rather than card billing. Procurement-friendly.

Custom term lengths. 1-year, 2-year, 3-year contracts with corresponding discounts. Multi-year terms typically get 10–25% off the per-year rate.

True-up provisions. If usage exceeds the contracted pool, overage bills at agreed rates. If usage falls below, some contracts roll forward unused credits; others let them lapse.

The negotiation is a normal enterprise sales process. Engage Cursor's account executive when you're ready to commit; they'll structure terms appropriate to your usage profile and risk tolerance.

A Compliance Review Checklist

For organizations preparing to deploy Cursor under formal compliance constraints, the checklist:

Data handling. - [ ] Privacy mode enforced org-wide. - [ ] DPA signed with Cursor. - [ ] Sub-processor list reviewed and approved. - [ ] Geographic data residency confirmed (if required). - [ ] Self-hosted indexing evaluated (if required).

Access control. - [ ] SCIM provisioning configured. - [ ] Role-based access matched to IdP groups. - [ ] Contractor scoping defined. - [ ] Admin role assignments deliberate (not default).

Operational. - [ ] Audit log ingestion configured. - [ ] SIEM integration tested. - [ ] Incident response runbook covers Cursor incidents. - [ ] On-call rotation knows how to triage Cursor issues.

Configuration governance. - [ ] Org-wide rules library defined. - [ ] Repository blocklists set for sensitive code. - [ ] Model blocklists set if relevant. - [ ] MCP allowlist defined (curated catalog only, or custom-approved).

Compliance specifics. - [ ] SOC 2 Type 2 report reviewed. - [ ] Penetration test results reviewed. - [ ] Encryption posture verified. - [ ] Incident notification SLAs documented. - [ ] Insurance coverage verified (cyber liability, etc.).

Engineer enablement. - [ ] Org-wide CONTRIBUTING.md or similar covering Cursor usage. - [ ] Onboarding flow established. - [ ] Champion identified per team to handle Cursor questions. - [ ] Budget alerts configured.

The checklist isn't exhaustive but covers what most security reviews ask for. Adapt it to your organization's specific framework — HIPAA-aligned organizations add HIPAA-specific items, FedRAMP-adjacent organizations add FedRAMP-specific ones.

Operational Practices at Scale

A few practices that distinguish well-run Enterprise Cursor deployments from struggling ones:

A dedicated owner. One person — usually a senior engineer or engineering manager — owns the org's Cursor configuration. They run the quarterly audits, review the analytics, evaluate new features, and serve as the contact point for Cursor's account team. Without an owner, configuration drifts and the deployment degrades.

A champions network. One champion per team. Champions are the local Cursor expert who handles team-level questions and feeds patterns back to the central owner. The network distributes knowledge without creating bottlenecks.

Quarterly business reviews with Cursor. Cursor's enterprise team holds QBRs covering usage trends, new features, the customer's concerns and successes. Useful for catching issues early and influencing the product roadmap.

A feedback loop with engineering. Engineers should have a clear path to surface friction (Slack channel, doc, regular survey). Pain points that don't surface stay pain points. The owner hears the feedback and either addresses it locally or escalates to Cursor.

Periodic re-evaluation. Every year or two, re-evaluate the deployment. Has the product changed in ways that warrant new configuration? Are the rules still right? Are the MCP integrations still useful? Maintenance is light if done regularly; it's a major project if deferred for years.

When Enterprise Is the Wrong Fit

Some signs Enterprise isn't the right answer:

You're a 20-person engineering team without formal compliance requirements. Teams plan is sufficient. Enterprise's overhead isn't worth the marginal capabilities you'd actually use.

Your security team is hostile to all AI tooling. Enterprise won't unblock that. The block is cultural, not contractual. Solving it takes education and stakeholder work, not a different Cursor SKU.

Your organization is in a regulated context that genuinely can't approve any hosted AI. Some defense, intelligence, and certain healthcare contexts fit. Cursor isn't approvable in those contexts; self-hosted alternatives or on-prem deployments are the path.

You haven't established the operational maturity Enterprise expects. Enterprise contracts assume a certain level of organizational support — audit log ingestion, owner assignment, lifecycle automation. Without these, the contract's overhead doesn't pay off.

The honest framing: Enterprise is the right tier for organizations that need it. Most don't, and Teams or Pro is the better fit. Honestly assess what your organization actually requires.

Part VIII: Safety, Ethics, and Limitations

"The most useful documentation is the part that tells you where the tool fails."

— A working principle

Chapter 29: What Cursor Gets Wrong

This is the chapter most product books don't write. Cursor is genuinely good — the rest of the book has been about getting the most out of it. It's also genuinely flawed, in ways that recur consistently enough to deserve a chapter. Knowing the failure modes makes you faster at spotting them, recovering from them, and avoiding the work patterns that trigger them.

The chapter is opinionated and specific. The failure modes below are the ones I see most often in working teams' Cursor use. Yours may differ; the categories are the durable part.

Hallucinated APIs

The most common failure mode across all AI editors, Cursor included: the model proposes code that calls a function, method, library, or API that doesn't exist. Often it looks plausible — the function name is reasonable, the arguments look right, the return type is what you'd expect. Run the code, and the function isn't there.

Hallucinated APIs come from a few patterns:

Library version drift. The model's training data includes the library at one version; your project uses another. The model proposes the API that existed in version 4; you're on version 5, where it was removed. Especially common after major version bumps.

Cross-language confusion. Functions that exist in Python's version of a library don't exist in JavaScript's version, and vice versa. The model occasionally blends them. Your `fetch().then()` chain ends up with a `.text()` method that looks like Python's.

Plausible-but-fictional convenience APIs. "There's totally a `User.findByEmail()` method" — except there isn't, but the model thinks there should be, so it writes one. The pattern matches what other ORMs offer; your specific ORM doesn't.

Hallucinated transitive dependencies. "Just use the `quick-validator` package." That package doesn't exist; the model invented it.

The frequency of this failure has dropped substantially in the 2024–2026 era as models improved, but it's not zero. Even Opus 4.7 hallucinates APIs occasionally on unfamiliar libraries.

What to do about it:

Run things. The fastest way to catch a hallucinated API is to run the code. Type checking catches some; running catches all of them. Don't merge code you haven't run.

Don't accept code you can't verify. If the model proposes `userService.findByEmailWithCache(email)` and you've never seen that method, look it up before accepting. Either it exists (good — accept) or it doesn't (don't).

Project rules pointing at the canonical APIs. "User lookup uses `userService.findById(id)` only. There is no `findByEmail`; the email-based lookup goes through `authService.userByEmail(email)`." A rule that captures the canonical API stops the model from inventing nearby alternatives.

Be more careful in unfamiliar libraries. When you're using a library you don't know well, the model is likelier to hallucinate. Read the proposed code with extra attention; consult the docs when something looks novel.

Long-Session Quality Decay

A second pattern, distinct from hallucination: as a chat session grows long, the quality of the model's output degrades. Early in the session, suggestions are sharp. By message thirty or forty, suggestions wander — referencing earlier messages incorrectly, proposing changes that contradict prior decisions, getting confused about what's already been done.

The cause is the model's working context filling with conversation history. As the context approaches the model's effective limit, attention spreads thin, older messages get summarized or evicted, and the model loses thread continuity.

The decay varies by model. Long-context models (Gemini 3.1 Pro at 1M+) hold up longer than shorter-context ones. Reasoning-focused models hold up better than fast/cheap ones. But all of them eventually degrade.

What to do:

Commit progress frequently and start fresh sessions. When a session has been running for an hour and the suggestions are turning fuzzy, commit what you have and open a new chat. The fresh session has clean context.

Notice the symptoms. Suggestions referencing things "we discussed earlier" but getting them wrong. Proposed changes that contradict an earlier decision. The model agreeing with everything you say (a sign it's not engaging carefully). When you see these, end the session.

Use Background Agent for genuinely long tasks. Background Agent has its own context lifecycle independent of your editor session. A long autonomous task in Background Agent doesn't degrade the way a long foreground chat does (different context profile).

Don't try to "remind" the model of earlier context. Re-pasting summaries into a long session rarely works; it just adds more context to a context that's already overloaded. Better to start fresh with a tighter scope.

Context Fragmentation in Big Agents

A specific case of long-session decay that hits Agent mode hard. Agent's loop fills its context with file reads, command outputs, partial work, and planning notes. Within a single Agent run, context grows fast — sometimes a 30-minute Agent run accumulates more context than a four-hour chat session.

The symptom: Agent's later actions in a run are less coherent than its earlier ones. It might re-read files it just edited, lose track of which tests it already ran, propose changes that conflict with changes it made twenty minutes ago.

What to do:

Scope Agent runs tighter. A focused 10-minute Agent run on a clear task degrades less than a sprawling 60-minute run on a vague one. The cure is mostly preventive — don't let Agent runs grow large to begin with.

Commit and restart for multi-stage work. A complex feature might need three Agent rounds: scaffolding, implementation, polish. Commit between rounds and start fresh contexts. Each Agent run sees only what it needs.

Use long-context models for inherently long tasks. Gemini 3.1 Pro for an Agent task that genuinely needs to read many files. Don't make Opus 4.7 (with shorter context) struggle on a task that fits Gemini's strengths.

Watch for the symptoms. Same as long-session decay — Agent referencing prior actions incorrectly, proposing redundant changes, going in circles. When you see these, stop the run.

Monorepo Cross-Package Leakage

A specific failure pattern in monorepos: the model edits package A but the changes leak conventions from package B (where you weren't editing). The packages have different styles, testing patterns, and error-handling conventions, and the model blends them.

Common shapes:

- The frontend package (React, hooks, JSX) gets a class-based change because the model saw classes in the backend package.
- The backend package gets ES modules syntax when the project uses CommonJS, because the frontend package uses ES modules.
- A test file in package A uses Jest patterns when the package actually uses Vitest, because Jest patterns are dominant in the rest of the monorepo.

The model isn't doing this maliciously; it's pattern-matching across the available context, and the codebase index gives it visibility into all packages at once.

What to do:

Per-package rules. `.cursor/rules/frontend.mdc`, `.cursor/rules/backend.mdc`, etc., each with glob scopes that limit them to their package. The rules tell the model what conventions apply where.

Tighter selections in Cmd+K and Manual. When editing code in package A, attach package A's files (not package B's) to the context. The model can still pull from the index, but you've signaled where the focus belongs.

`.cursorignore` strategically. If two packages keep getting confused, exclude one from indexing while you work on the other. Heavy-handed but effective when the leakage persists.

Project structure that helps. If the monorepo's per-package conventions diverge sharply, surface that in the `README.md` and in rule files. The more salient the differences, the less the model blurs them.

The 5,000-Line File Slowdown

Mentioned in earlier chapters but worth flagging here as a failure mode: very large individual files cause Cursor's quality to degrade specifically for code in those files.

The mechanism: Cursor's embedding pipeline summarizes files past a certain size threshold rather than embedding them in full. For files past 5,000 lines, the model's grasp of the file's contents is shallower than for normal-sized files. Tab autocomplete is weaker. Cmd+K rewrites are less precise. Multi-file edits that include the large file lose precision.

What to do:

Split the file. If you can. A 5,000-line file usually wants to be a directory of focused 200-line files anyway. The split improves both Cursor's understanding and human review-ability.

Accept the degraded experience. If splitting isn't feasible (generated code, framework-imposed structure), Cursor's behavior on that specific file will be weaker than elsewhere. Use Manual mode with the file explicitly attached so the model reads it directly.

Exclude from indexing. If the file is generated, ignore it. The index doesn't help when the file is regenerated on every build.

Edit smaller files when possible. Sometimes work that "needs to touch the 5,000-line file" can be done by adding a smaller file that the large one imports. Architectural workaround for the indexing limitation.

The Confidence Trap

A subtle failure mode: the model writes confidently regardless of how confident it should be. There's no calibrated uncertainty — the model's "this is the right approach" sounds the same when it's right and when it's wrong.

This shows up everywhere:

- Tab completions look identical whether the model's prediction is high- or low-confidence.
- Chat responses use the same authoritative tone for "I know this with high confidence" and "I'm guessing."
- Agent reports "task complete" with the same affect when it actually completed and when it papered over failures.
- Manual proposes diffs that look polished even when the underlying logic is wrong.

The user-side discipline: treat the model's confidence as uninformative. Don't accept code because "the model seems confident." Verify directly. Confidence isn't a signal you can use.

Workaround Spirals

Specific to Agent mode: when the model hits an obstacle (failing test, missing dependency, unexpected error), its impulse is often to work around the obstacle rather than diagnose it. Round one: a workaround for the test failure. Round two: a workaround for the new test failure introduced by round one. Round three: increasingly complicated code to handle the cumulative tangle.

By round four or five, the diff is unrecognizable. The "fix" is layers of complexity that wouldn't have been necessary if the model had stopped to diagnose at round one.

What to do:

Stop spiraling Agent runs. When you see Agent making the same kind of fix repeatedly without progress, intervene. Pause the run, look at the actual problem, and either give Agent the diagnosis or take over manually.

Set explicit constraints. "If a test is failing, diagnose the root cause before making any code changes. Don't change the test itself unless we've confirmed the test is wrong." The constraint discourages workaround-first reasoning.

Switch to Ask mode when you suspect a spiral. Ask the model to diagnose the failure, identify the root cause, and propose the right fix. Then implement deliberately.

Confident Ignorance of Newer Conventions

Models' training data has a cutoff. APIs, libraries, and patterns that became standard after the cutoff aren't in the model's knowledge by default. The model still proposes patterns confidently — they're just outdated patterns.

Examples:

- A new React feature (Server Actions, useTransition variants, etc.) released after the model's training. The model proposes the older patterns it knows.
- A library's recommended pattern changed (TanStack Query's new query options API, for example). The model proposes the deprecated form.
- New TypeScript features (the `satisfies` operator, decorators stage 3, etc.) that aren't yet reflected in training.

What to do:

@Web search for current best practices. When working with bleeding-edge APIs, attach @Web so the model can consult current docs.

Project rules that explicitly call out the current pattern. If your project uses a new pattern, document it in rules. The model reads the rules and aligns.

Be skeptical of "use this approach" suggestions for fast-moving libraries. Verify against current docs.

Update your model. Cursor's picker offers newer versions as they ship. Keep your default current.

The Drift Across Sessions

A team-scale failure mode: rules drift across sessions. Engineer A asks the model to do X; the model does X but takes some liberties. Engineer A doesn't notice. Engineer B asks the model to do something similar later; the model has been retrained or updated, and now it does X slightly differently. Engineer B's code looks slightly different from Engineer A's.

Over months, the codebase accumulates subtle convention drift driven by per-session model variations. No individual change looks wrong; the cumulative pattern is messy.

The cure is the same as for any convention drift: rules. Project rules are the pin that prevents drift. Without them, expect drift; with them, the model anchors on the project's specific conventions across sessions.

Failure-Mode Taxonomy Summary

A condensed table of the failure modes covered in this chapter:

Failure	Trigger	Detection	Mitigation
Hallucinated API	Unfamiliar library or version drift	Code doesn't compile/run	Rules pointing at canonical APIs; verify before accepting
Long-session decay	Chat sessions over ~1 hour	Vague responses, contradictions	Commit and start fresh sessions
Context fragmentation	Long Agent runs	Agent goes in circles	Tighter Agent scope; long-context models for long tasks
Monorepo leakage	Cross-package context confusion	Style mixing between packages	Per-package rules with glob scoping
5,000-line file slowdown	Files past size threshold	Weaker Tab/Cmd+K in big files	Split files; explicit attachment for big-file edits
Confidence trap	All Cursor output	Look for verification before acceptance	Don't trust confidence; verify always

Failure	Trigger	Detection	Mitigation
Workaround spirals	Agent hitting obstacles	Repeated similar fixes, no progress	Stop spiraling Agent runs; constrain to diagnose-first
Outdated conventions	Newer APIs post-training cutoff	Proposed code uses deprecated patterns	@Web for current docs; rules for current patterns
Cross-session drift	Variable model behavior over time	Codebase style accumulates inconsistency	Strong project rules anchor the model

From Failure Modes to Trust

The failure modes are real. The book isn't pretending Cursor is perfect; it's showing you where to look so the failures don't surprise you. Chapter 30 zooms out from specific failure modes to the broader question — how do you trust AI-generated code at all? The answer is the discipline of code review: treating every AI-generated change with the same scrutiny you'd give a junior engineer's first PR, every time.

Chapter 30: Code Review and Trust

The most important skill in AI-assisted development isn't writing prompts. It's reading code. Specifically: reading every line of AI-generated code as carefully as you'd read code from a junior engineer you've decided to bet on but who occasionally makes subtle mistakes. The discipline is "every line, always" — the phrase that's appeared in earlier chapters and that anchors this one.

This chapter is about why the discipline matters, what good AI-code review looks like, and the patterns that protect you from trusting too much.

The Trust Question

When the model proposes a diff, you have two choices. Read it carefully and decide whether to accept, or skim and trust.

The case for skimming: the model is right most of the time. Reading every line takes time. Speed is part of why you're using Cursor at all. Treating every diff as suspect feels like undermining the tool's value.

The case for reading: the model is wrong some of the time. The wrong cases include subtle bugs that look fine at a glance. The cost of merging a subtle bug to production is much higher than the cost of reading carefully. The "speed" you gain from skimming is borrowed against future debugging time.

The math favors reading carefully, by a wide margin. A senior engineer's experience: the bugs that ship to production aren't the obvious ones; they're the ones that look fine. The reading discipline is what catches the not-fine ones.

Why "Every Line, Always" Is Non-Negotiable

The phrase "every line, always" sounds dogmatic. It is, deliberately. The discipline relaxes only in two cases:

1. **The diff is reverting your previous change.** You wrote it ten minutes ago. You know what's there.
2. **The diff is auto-formatting your code.** Prettier ran; you trust it. You can scan rather than read.

Otherwise — for every Tab proposal, every Cmd+K result, every Manual diff, every Agent step, every BugBot fix PR — read every line.

The reasons this is non-negotiable:

The model's confidence isn't calibrated. Confident output looks like confident output, whether the underlying code is right or wrong. Reading is the only way to distinguish.

Bugs cluster at the boundaries. The lines you'd expect to be most carefully written (the model's "main idea") are usually right. The bugs cluster at the edges — error paths, boundary conditions, less-trafficked branches. Skimming catches the obvious; reading catches the rest.

The cost is asymmetric. A few seconds of reading vs. a production bug. The math compounds in favor of reading even at low base-rate failure rates.

Habit beats heroics. A team that reads every line consistently has fewer production incidents than a team that reads carefully only on "important" PRs. Defining "important" is harder than reading every line.

Reading builds your model of the model. Over time, reading every line teaches you when the model is reliable and when it isn't. The intuition for "this kind of task the model gets right; this kind it gets wrong" comes from reading. Skip the reading and you skip the learning.

The Reading Checklist

A condensed version of the checklist from Chapter 12, adapted to the broader review case:

Does the diff do what was asked? Match the request to the result.

Does the diff do anything that wasn't asked? Watch for "while I was here" cleanup.

Are the boundaries right? First and last lines of the diff often hide off-by-one errors.

Are the conditions right? Read every `if`, ?, &&, || carefully. Polarity flips are subtle and devastating.

Are error paths preserved? Refactors often weaken error handling without flagging it.

Are names accurate? Renamed-but-different variables, shadowing, typos in identifiers.

Did dependencies change? New imports, version bumps, package additions. Each is a flag.

Are there TODOs or placeholders? Models sometimes leave incomplete sections.

Are tests updated correspondingly? Behavior changes without test changes are flags.

Does the change make sense in context? Sometimes the diff is locally correct but globally wrong — fits the request but not the system.

Run the checklist on every diff. With practice, it collapses into one mental pass that takes seconds, not minutes.

Reading at Different Scales

The reading discipline shows up at different scales depending on the mode:

Tab proposals. A glance, a half-second of attention. Match against what you would have typed.

Cmd+K diffs. A few seconds. The diff is small; the read is small.

Manual edit diffs. Per file, per hunk, per line. A full Manual edit might take 1–3 minutes of careful reading depending on size.

Agent runs. Watch as it works (catch direction issues early). Read the final cumulative diff. Run the verification commands yourself. 5–10 minutes of post-run review for a substantial Agent task.

Background Agent PRs. Treat like any junior engineer's PR. Full code review. Possibly 10–20 minutes for a meaningful PR.

BugBot fix PRs. Full code review. The fix may or may not be correct.

Multitask outputs. Comparison plus review. 10–30 minutes total to compare and pick.

Reading time is part of the workflow, not overhead. Budget for it.

When the Model Is Wrong

A few specific patterns where the model is wrong reliably enough to flag:

Inverted polarity. "Replace `foo === true`" with "`bar === false`" — the model gets the rename right and the polarity flip wrong. Especially common when the rename and the polarity flip are conceptually linked but mechanically separate.

Missing edge cases in test "improvements." "Improve this test" sometimes means the model adds a happy-path case and drops the edge case it didn't fully understand. Read the new tests against the old ones.

Subtly different behavior in a "preserving" refactor. "Refactor without changing behavior" — the model claims it preserved behavior but actually changed something subtle. Diff the new function's behavior against the old one mentally; don't trust the model's claim.

Off-by-one in boundary conditions. "Loop until N items" vs. "loop while less than N items" produces a different boundary. The model occasionally gets it wrong.

Silently swallowed errors. A refactor that catches an error and doesn't re-throw, or handles a Promise rejection that used to propagate. The behavior change is invisible until production.

Dependency version drift. "Add this library" — the model picks a version. Sometimes the pick is wrong (very old, recently yanked, has known security issues). Verify the version.

Removed or weakened security checks. The model occasionally rewrites a function and "simplifies" the auth check away. Always read security-adjacent code with extra attention.

When you spot any of these in your own work, the lesson is: the model wrote it confidently. Confidence is uninformative. Reading caught it.

Test Discipline in the Agent Era

Tests are your verification layer. With agents producing diffs faster than humans can write them, tests become more load-bearing, not less.

Some patterns:

Don't trust "tests pass" claims from Agent. Run the tests yourself before merging. The exit code is the truth; the model's report is text.

Add tests for behavior the model can't verify. Anything where a test would catch a subtle bug — boundary conditions, error paths, security checks — write the test. The cost is small; the payoff when the test catches a model-introduced bug is many times over.

Run the full suite before merging. Not just the changed tests; the full suite. Cumulative changes have cumulative effects.

Use a CI pipeline that's strict about coverage and regressions. A CI that catches "tests deleted in this PR" or "coverage dropped 10% in this PR" backstops the model's occasional test-deleting tendencies.

Watch for tests that pass for the wrong reason. A test that passes because the assertion is too weak (it asserts something true regardless of behavior) isn't a real test. The model occasionally writes these without realizing.

Who Owns AI-Written Code?

A question that's increasingly asked: who's responsible for code the AI wrote?

The answer is the human who accepted the change. The model proposed; the human accepted. Accountability is the human's.

The implications:

You can't blame the AI. "The AI wrote it" isn't a defense for a production bug. You accepted the diff; you own the bug.

Code review approves AI-generated code the same as human-generated code. Reviewer signs off; reviewer is accountable for catches and misses.

Audit logs and AI-tracking metadata aren't a way to assign blame elsewhere. They're tooling for understanding what happened. Accountability remains with the engineer.

This is fine. The framing is the same as it's always been for code review. AI changes who does the typing; it doesn't change who's responsible.

The cultural shift in some teams: engineers who lean heavily on AI need to internalize that they're still responsible for the output. "I just hit Tab" isn't a defense. Accountability scales with the volume of AI-assisted work.

The Trust Calibration

A subtle but important distinction: trust in the tool vs. trust in any particular output.

You can trust Cursor as a tool. It's well-built, it's reliable, and it isn't maliciously trying to mislead you. The base-rate quality is high.

You can't trust any particular output without verification. The base-rate quality means most outputs are good; the failure rate is low but not zero. Your job is to filter the no-good outputs from the good ones, every time, on every output.

This calibration matters. Distrusting the tool entirely (refusing to use Cursor) sacrifices the leverage. Trusting it blindly (accepting all outputs) ships the bugs. The middle path — trust the tool, verify each output — is the one that actually works.

Building the Habit

Reading every line is a habit, not a rule you follow consciously. Three practices build the habit:

Read out loud at first. Literally read the code aloud as you review it. Slows you down; forces attention. After a few weeks, drop the out-loud part; the attention pattern remains.

Pair-program with yourself for the first month. When the model proposes a diff, pretend a colleague is sitting next to you and you have to explain the diff to them before accepting. Articulating what's happening forces engagement.

Review your accept/reject ratio. If you've accepted 100% of the model's proposals in a session, you're not reading carefully — even the best model has some wrong outputs. A healthy reading discipline produces a meaningful reject rate. Notice if yours is too high.

After a month or two, the habit goes automatic. You stop noticing that you're reading carefully; the reading is just part of how you accept code.

The Model Is Not Your Reviewer

A clarifying point: when the model is the one writing the code, the model can't also be the one reviewing it. You're the reviewer.

Some teams blur this — using BugBot to review BugBot's fix PRs, or asking Agent to "double-check the changes you just made." These are useful diagnostic steps but they aren't review. The model checking its own work has the same blind spots that produced the work in the first place.

Real review needs different eyes. Either yours or another engineer's. BugBot can be a first pass; it isn't the review.

For solo developers, this means you're always the reviewer. There's no one else. The discipline matters most when no one else is checking.

The failure modes (Chapter 29) and the review discipline are in place. The remaining hard topics are security — what AI editors mean for code-level and operational security — and ethics. Chapter 31 takes the security side: threat model for using Cursor on sensitive code, and operational practices that hold up under scrutiny.

Chapter 31: Security and Compliance

Security and compliance get treated as enterprise concerns, but they apply at every scale. A solo developer working on client code has a security obligation. A small team handling user data has a compliance obligation. An enterprise has both, with paperwork. This chapter is about what using Cursor responsibly looks like across that range — the threat model, the controls, and the operational practices that hold up under review.

The chapter complements rather than repeats Chapters 21 (Privacy mode) and 28 (Enterprise). The focus here is the integrated picture: how the pieces fit together for organizations that can't afford to treat security as an afterthought.

What Happens to Your Code

The first thing to understand is the data flow. When you use Cursor, your code touches several systems:

Your local machine. The editor runs locally. Files live on your disk. Standard local-machine security applies — disk encryption, OS updates, antivirus, screen lock, etc. None of this is Cursor-specific; it's general developer-laptop hygiene.

Cursor's infrastructure. When you invoke an AI feature, code goes to Cursor's servers for context retrieval, indexing, and routing to model providers. Privacy mode controls retention; in-transit handling happens regardless of the setting.

Model providers. Anthropic, OpenAI, DeepSeek, Google. Cursor relays your context to whichever model you've selected. The provider sees the data during inference. Each provider has its own data handling — typically zero-data-retention for paid API tiers, but verify per-provider for your specifics.

MCP servers. When the model invokes an MCP tool, the tool's inputs and outputs cross Cursor's infrastructure. The MCP server itself may contact external systems (Linear, Sentry, etc.); those systems see what they see per their own policies.

Background Agent infrastructure. Background Agent runs on Cursor's cloud sandbox. The sandbox clones your code, runs commands, and produces output. It's isolated per-tenant, but the data flow is "your code goes to Cursor's cloud."

External integrations. GitHub for PR creation, Slack for notifications, your IdP for SSO, your SIEM for audit logs. Each is a system that sees data per its own role.

A complete data-flow diagram for your specific deployment is the artifact most security teams need. Cursor's enterprise team will provide the canonical version on request; tailoring it to your configuration is your work.

The Threat Model

The risks worth thinking about, ordered roughly by likelihood:

Accidental disclosure of sensitive data through normal use. Highest-frequency risk. An engineer asks the model about a secret-laden file. Even with Privacy mode, the data was processed in transit. Mitigation: `.cursorignore` for sensitive files; never let the editor see what shouldn't be seen.

Insider risk at Cursor or providers. A Cursor or model-provider employee accesses customer data they shouldn't. Lowered by Privacy mode (less data exists to access) and by the providers' internal access controls. Real but bounded.

Breach at Cursor. A security incident at Cursor exposes whatever Cursor has retained. Lowered by Privacy mode (less is retained). Real but narrower in impact than for many SaaS tools.

Legal compulsion. Subpoena or government request. Cursor turns over what it has. Privacy mode means less to turn over. For logging-based requests, depends on Cursor's logging retention.

Compromise via MCP integration. A poorly vetted MCP server has security issues — leaked credentials, malicious behavior, supply chain compromise. Mitigation: stick to the catalog or vet custom MCP servers before installing.

Agent-driven action causing damage. Agent runs a destructive command (rm, drop database, push –force). Mitigation: per-command approval discipline, explicit blocklist of dangerous commands, restricted privileges where Agent runs.

Model-injected vulnerabilities. The model writes code with a subtle security flaw (SQL injection, XSS, missing auth). Mitigation: code review, security-focused project rules, BugBot or equivalent automated scanning.

Supply chain attacks via model output. The model proposes a dependency that turns out to be malicious. Mitigation: review dependencies the model adds; apply the same supply-chain scrutiny you'd use for any dependency.

The threat model covers each of these with specific mitigations. Most are addressed by good defaults plus operational discipline; a few require organizational policy.

Privacy Mode Threat Model

Specifically focused on Privacy mode's role in the picture:

Privacy mode protects against retention-based risks. Breach, legal compulsion, insider access — all benefit from Cursor having less retained data. Privacy mode is the primary lever.

Privacy mode does not protect against in-transit visibility. The model provider sees your code during inference. Cursor's servers see it during routing. Privacy mode doesn't change visibility during processing.

Privacy mode does not protect against MCP-mediated leaks. When Agent calls an MCP tool, the tool's inputs go through Cursor and out to the integrated system. That data flow runs independent of Privacy mode.

Privacy mode does not protect against accidentally putting secrets in code. If you commit credentials and let Cursor index them, they're processed regardless of Privacy mode.

For most teams, Privacy mode plus reasonable hygiene suffices. For higher-risk teams, additional layers (BYOK, self-hosted models, restricted MCP, repo blocklists) close more of the gap.

Secrets in Agent Transcripts

A specific failure pattern: secrets ending up in chat history or Agent transcripts.

The pattern: an engineer is debugging an authentication issue. They paste a token into chat to ask "why doesn't this work?" The token now lives in the chat history. Chat history is preserved per-project for the team to consult later. The token is now in any backups or logs that touch the chat history.

What to do:

Don't paste secrets into chat. Ever. Refer to secrets by name ("the production database password"), not by value. If the model needs to see the secret format, paste a redacted version.

Use environment variable references. "$DATABASE_URL" instead of the URL. The model can reason about the variable name without seeing the value.

Audit chat history periodically. If a team relies on long-running chat, audit for accidental secret pastes. Delete what shouldn't be there.

Train engineers on the failure mode. New engineers don't know to avoid pasting secrets into chat. Add it to onboarding. The cost of training is low; the cost of a leaked production token is high.

Configure secret scanners. If your repo has secret-scanning enabled, the same protections apply to commits the model produces. Catch the leak at commit time as a backstop.

Working Under HIPAA

Healthcare-adjacent organizations have specific HIPAA requirements that interact with AI tooling:

No PHI in prompts. Patient health information should not be sent to AI providers. This is structural — even with Privacy mode and BYOK, sending PHI to an AI model is a HIPAA-violating disclosure.

Code about PHI is fine; PHI itself isn't. A function that processes patient records can be reviewed by AI; an actual patient's record cannot.

Test data must be synthetic. Don't use real patient data in test fixtures, even anonymized. Use synthetic data the AI can see freely.

Review for inadvertent PHI handling. Sometimes a comment or test mentions a real patient. Audit for this, and `.cursorignore` files that contain PHI.

Document the boundary. Your HIPAA documentation should explicitly cover AI tooling — what's allowed, what's not, what controls are in place. Engineers should know what they can and can't ask the model.

For HIPAA-bound organizations, the realistic path is: AI tooling for code review and code generation; no PHI ever enters the AI flow; clear engineer education about the boundary. With that discipline, Cursor is usable. Without it, you're risking a HIPAA violation per AI invocation.

Working Under SOC 2

SOC 2-aligned organizations have a different shape of requirements — controls around access management, change management, audit logging, vendor management. Cursor's role:

Vendor due diligence. Cursor must be on the vendor list with documented review. Cursor's SOC 2 Type 2 attestation supports this.

Access management. SCIM provisioning ties Cursor access to your IdP, satisfying access-control requirements.

Change management. Cursor's role-based controls let you specify who can change what. Audit logs document the changes.

Logging and monitoring. Audit logs flowing to your SIEM satisfy logging requirements. Monitor for anomalies (usage spikes, unexpected configuration changes).

Encryption. Cursor's in-transit and at-rest encryption is documented; verify it meets your standards.

Incident response. Cursor has documented incident response procedures; make sure your runbook covers Cursor incidents.

For SOC 2, the work is mostly process: documenting that Cursor is in your environment, that you've done the due diligence, that the controls are in place, that you monitor for anomalies. The technical posture is mostly already there; the documentation work is yours.

Working Under FedRAMP-Adjacent Requirements

Federal-government-adjacent work — FedRAMP, ITAR-aligned, certain DoD contexts — typically has stricter requirements than SOC 2 or HIPAA. Cursor's posture:

FedRAMP authorization. Cursor's posture varies; check current status. Many SaaS tools aren't FedRAMP-authorized; this is a hard limit if your contracts require it.

Data residency. Cursor's enterprise contracts can specify geographic data residency. Verify the regions match your requirements (US-only, US Federal cloud, etc.).

ITAR and export-controlled data. ITAR-controlled code cannot be processed through services that may have non-US-person access. A structural constraint that often rules out hosted AI tools entirely.

Air-gapped environments. Some federal environments are air-gapped. Cursor's hosted model doesn't fit. Self-hosted alternatives (custom builds of open-source AI editors with self-hosted models) are the path.

For federal-adjacent work, the realistic answer is often "Cursor doesn't fit, use a different tool" or "Cursor fits for the unclassified portions of work." Don't try to force-fit Cursor into a regime it can't satisfy; the consequences of getting compliance wrong are severe.

The Enterprise Compliance Overlay

Enterprise contracts add the contractual machinery for serious compliance work:

Custom DPAs. Data Processing Agreements specifying exactly what data is processed, retained, and how long.

Sub-processor agreements. Each model provider Cursor uses is a sub-processor; the contract chains through.

Custom data residency. Specific geographic processing requirements.

Custom retention policies. Beyond Privacy mode's structural defaults, contractual commitments to retention behavior.

Audit rights. Some enterprise contracts include the customer's right to audit Cursor's compliance posture.

Incident notification SLAs. Specific timelines for Cursor to notify the customer of security incidents affecting their data.

Liability caps. Contractual financial protections in case of incidents.

The contractual layer is what makes Cursor approvable in compliance-heavy contexts. None of it changes the technical behavior; all of it shapes the legal recourse and the documentation supporting compliance attestation.

Operational Security Practices

Beyond the contractual and configurational, daily security practices that matter:

Don't run privileged commands through Agent without thought. sudo, chmod, anything touching production systems, anything that escalates privilege. Per-command approval is your friend; auto-approving privileged operations is the failure mode.

Audit MCP installations. Each MCP server you install is code that runs in your environment. Vet them. Custom MCP servers should be code-reviewed before deployment. Catalog servers are vetted by Cursor; custom ones are your responsibility.

Watch for credential leaks in Agent output. Sometimes Agent's command output includes credentials (database connection strings, API keys, tokens). These end up in Cursor's session state and potentially in your audit logs. Configure your tools to redact credentials in their output where possible.

Secrets management for hooks and slash commands. Hooks can read environment variables; slash commands can reference them. Don't put secrets directly in `.cursor/hooks/` or `.cursor/commands/` — use environment variables or your project's secrets management.

Limit what Background Agent can do. Background Agent runs in a sandbox, but the sandbox has authority — it pushes branches, opens PRs, calls external APIs. Configure the sandbox's permissions deliberately.

Monitor unusual usage. A sudden spike in MCP tool calls, an Agent run accessing files outside its normal scope, a credential rotation that didn't propagate — these signal problems. Monitoring tools (your SIEM, Cursor's audit logs) catch them.

Threat Modeling Cursor for Your Team

If you're advocating for Cursor in an organization with a formal security review, the artifacts you'll likely need:

- Data flow diagram (Cursor's enterprise team provides a starting version)
- Sub-processor list
- Privacy mode posture documentation
- Compliance attestations (SOC 2, etc.)
- Incident response procedures
- Operational practices documentation (your team's policy on how Cursor is used)
- Risk assessment (the threat model applied to your specific use case)
- Mitigations (the controls you've put in place)

Most of these are produced once and maintained on a yearly cadence. Cursor's enterprise team has been through hundreds of these reviews; lean on their expertise rather than starting from scratch.

Chapter 32: The Ethics of AI-Assisted Development

The book has been mostly practical so far — how to use Cursor well, where it fails, how to be safe. The remaining piece is the harder, less-resolved set of questions about what AI-assisted development means for the field and for the people in it. Disclosure norms. License compatibility. Skill atrophy. The widening gap between engineers who lean on AI well and those who lean on it badly. The trajectory of where this all goes.

This chapter doesn't have clean answers. The field is too new, the practices too unsettled, the tradeoffs too contested. What the chapter can do is map the questions clearly enough that you can think about them deliberately rather than letting your practice settle by accident.

Disclosure: When and How

The question: when you ship code, do you disclose that AI helped write it?

The honest answer: norms are emerging and vary by context. A few cases where consensus is clearer:

Open-source contributions. Many open-source projects now have explicit policies on AI-assisted contributions. Some accept them with disclosure ("AI-assisted using Cursor"), some accept them without disclosure if the human reviewed carefully, some prefer no AI contributions at all. Read the project's contributing guide before submitting; respect what the project asks.

Commercial work for clients. If you're a contractor writing code for a client, the client's expectations matter more than your preferences. Some clients are comfortable with AI-assisted work; some aren't; some require disclosure. Ask. The question is professional and reasonable.

Academic and research contexts. Honest disclosure is expected. AI-assisted code in research papers, theses, course assignments — disclose. Failing to disclose in academic contexts has been treated as an integrity violation in multiple recent cases.

Bug bounty submissions, CVE disclosures, security work. Disclosure helps reviewers calibrate. Most security researchers now disclose AI use as a matter of course.

Where consensus is less clear:

Internal company code. Most teams don't require disclosure of AI assistance for normal product work. The expectation is usually "you reviewed it; it's your code now." Some teams keep tracking metadata for analytics but don't surface it to humans.

Legal documents, contracts, policy documents. AI-assisted text in legal contexts has been challenged in courts. Outcomes vary. Conservative practice is to disclose.

Code that becomes part of a product you're claiming credit for. Personal portfolio projects, hackathon submissions, code you're publicizing as "look what I built." How much credit is yours when the AI did meaningful work is unsettled. Conservative practice is to be transparent about the role AI played.

The book's recommendation: when in doubt, disclose. The cost of disclosure is low; the cost of being caught not disclosing in a context where it mattered is high. Honest framing protects you.

License Compatibility for AI-Generated Code

The legal landscape around AI-generated code's license posture is unsettled. Some shapes of the question:

Did the model train on code with a particular license? Yes — most foundation models trained on large amounts of public code, licensed code included. Whether the model's outputs constitute derivative works of that training data has been litigated and remains partially unresolved. Different jurisdictions reach different conclusions.

Does the model "remember" specific code from training? Sometimes. The phenomenon is called memorization, and it's been documented. The model can occasionally reproduce verbatim chunks of code it saw during training. Frequency is low for normal use cases; the risk increases for prompts that closely resemble training data.

What does this mean for the code you ship? The honest answer is "consult your legal counsel." A few practical guidelines that conservative legal teams suggest:

- For internal use, the risk is bounded; standard code review catches the obvious cases.
- For products shipping under a permissive license (MIT, BSD), the risk is bounded by the model's typical output behavior plus your review.
- For products shipping under a copyleft license (GPL), be more careful — copyleft requirements interact with potentially-derivative-work concerns.
- For products where your customer requires specific license cleanliness (no GPL, no AGPL), be very careful — the AI might propose code that looks clean but is structurally similar to something the customer wouldn't allow.

What about the AI provider's terms? Cursor and the model providers each have terms about what you can and can't do with output. Read them. Most modern providers grant the user broad rights to model outputs, but the specifics matter for commercial use.

What about your client's terms? Contractor work often has IP assignment clauses. Some clients now have specific clauses about AI tooling — what's allowed, what's not, how disclosure works. Read your contracts.

For most working developers in 2026, the practical answer is: use the AI tools, review the output carefully, document the use where context warrants it, consult your legal team for high-stakes IP situations. Don't ignore the question; don't treat it as fully resolved either.

The Juniors-vs-Seniors Gap

A real and worsening dynamic: AI tools widen the gap between effective and ineffective engineers, and the widening hits juniors hardest.

The mechanism: an effective senior engineer uses AI as leverage. They know what they want, they read the output critically, they catch the mistakes, they ship better code faster. The AI multiplies their ability.

A junior engineer using AI poorly accepts what the model produces, misses the mistakes, ships code with subtle bugs, and doesn't develop the underlying skills (because the AI did the work for them). Over months, the junior plateaus or even regresses while the senior accelerates.

The gap shows up in:

- **Code review.** Senior engineers' AI-assisted PRs are clean and considered; junior engineers' are messy and require more review.
- **Debugging skill.** Engineers who let AI do their debugging don't develop the diagnostic instincts that come from struggling through bugs themselves.
- **Architectural judgment.** AI is poor at architecture; engineers who outsource it stop developing the skill.
- **Production reliability.** AI-assisted code from less-experienced engineers ships more incidents than the same engineers' non-AI-assisted code, because the bugs they would have caught during slow manual writing slip through fast AI-assisted writing.

The implications are uncomfortable. Junior engineers who use AI tooling well develop faster than their pre-AI peers. Junior engineers who use it poorly develop slower. How to teach junior engineers to use AI well is a question the field is still working out.

What seems to help:

Pair programming with seniors. Watching a senior engineer use Cursor (the modes they pick, the prompts they write, the diffs they reject) teaches more than any documentation. Make this part of onboarding.

Mandatory review without AI assistance occasionally. Some teams require engineers to review at least some PRs without AI help, to keep their critical-reading skills sharp.

Periodic non-AI work. Sprints where junior engineers don't use AI for specific kinds of work — debugging a tricky issue, designing an abstraction, writing initial tests. Skills atrophy with disuse; deliberate non-use prevents that.

Honest feedback in code review. "This PR shows the model's voice rather than yours" is hard to give but valuable. Engineers can't develop their own voice if reviewers don't notice when it's missing.

The book is bullish on AI-assisted development overall. It's also bearish on the trajectory if the gap-widening dynamic isn't addressed. Teams that don't deliberately invest in junior development end up stuck with very productive seniors and juniors who never graduate to seniors.

Skill Atrophy

Adjacent to the gap question: even senior engineers can lose skills they don't exercise.

The pattern: a senior engineer used to be excellent at, say, complex SQL queries. Cursor handles SQL well, so the engineer stops writing complex SQL by hand. Two years later, they can't write the queries themselves; they've offloaded the skill.

For most skills, this is fine — we offload to tools all the time. We don't compute logarithms by hand because calculators exist. We don't write assembly because compilers exist. Tools letting us move up a level of abstraction is the normal arc of the field.

Sometimes it's not fine. Skills worth retaining:

Debugging fundamentals. When AI fails, you need to debug yourself. Engineers who can't reason about their own systems are stuck when AI can't help.

Architectural judgment. AI is poor at architecture. Engineers who outsource architectural decisions to the AI ship architecturally messy systems and don't realize it.

Reading code. Critical reading is the skill that catches AI mistakes. Engineers who stop reading carefully accept more wrong code.

Writing code from scratch. When the AI is wrong or unavailable, the ability to write code from a blank file matters. Atrophy here is dangerous.

Mental simulation of program execution. "What does this code actually do" depends on practice. AI doing the simulation for you (debugger output, hypothesis chains) substitutes for the practice.

The hedge: don't fully outsource these skills. Periodic non-AI work, deliberate practice on the skills you want to retain, awareness of when you're letting the AI do something you should still do yourself.

The Long-Term Trajectory

A speculative section that ages worst, but worth attempting: where is AI-assisted development going?

Several trends seem durable:

The IDE becomes an agent runtime more than an editor. Cursor 2026 is most of the way there. The 2030 version will be further. The "editor" abstraction will be increasingly secondary to "AI workflows that happen to have an editor surface."

Cost curves flatten and access broadens. AI inference costs have dropped substantially over the past two years. The trajectory continues. The gap between "expensive AI" and "free AI" will keep narrowing, with implications for who can afford to use these tools.

Specialty models proliferate. General-purpose foundation models get supplemented by specialty models tuned for specific domains — security review, testing, documentation, refactoring. The picker grows, not shrinks.

Verification becomes the bottleneck. As AI generates more code, human work shifts from typing to verifying. The skills that matter shift accordingly. Reading carefully, designing systems for verifiability, building strong test suites — these become more valuable, not less.

Workflow discipline becomes the durable skill. The specific tools change. The discipline of scoping work, picking the right mode, reading every line, maintaining shared rules — that survives across tool transitions. Engineers who internalize the discipline ride waves of tool evolution; engineers who don't fight each new tool from scratch.

Auditability becomes mainstream. "What did the AI do, and when?" becomes a normal question infrastructure has to answer. Cursor's AI-tracking API and audit logs are early versions of what becomes table stakes.

The harder predictions are about second-order effects: what happens to junior developer pipelines, what happens to small-team economics, what happens to programming as a profession in five and ten years. The honest answer is "we don't know, but the trajectory is steep enough that it's worth watching."

What This Means for You

A few practical takeaways from the ethics chapter:

Disclose when context calls for it. Default to transparency in unfamiliar contexts. Read the conventions of the communities and clients you work with.

Document AI use in commercial work. Even if your client doesn't require it, internal documentation that says "we use Cursor; here's our policy on review and verification" protects you when questions come up later.

Invest in junior development. If you lead a team, the gap-widening dynamic is your problem to solve. Pair programming, deliberate non-AI work, honest feedback — these aren't optional if you want your team to grow.

Practice skills you want to retain. Skills you don't exercise atrophy. Choose deliberately what to keep practicing yourself, even when AI could do it.

Stay calibrated about tool change. The Cursor of 2026 isn't the Cursor of 2027. The discipline matters more than the specific tool. Build practices that survive tool transitions.

Take the ethics seriously. They're not a side topic. They shape the field's direction and your personal trajectory in it. Engineers who think about these questions deliberately develop differently than engineers who don't.

Where This All Goes

The book is a practical guide to a tool that, even in 2026, is still rapidly evolving. Specific features will change. The pricing model will shift again. The model picker will look different in two years. New capabilities will arrive that this book doesn't cover.

What endures: the workflow discipline. The reading habit. The mode rubric. The rules-driven approach. The skepticism of confident-looking AI output. The investment in shared team configuration. The honest assessment of where the tools fail.

If you read the rest of this book and remember nothing else, remember those. They're durable. The product on top is just the current incarnation.

Closing the Practical Chapters

You've finished the chapters on the limits and ethics of AI-assisted development. The book moves now into closing material — the Epilogue and the Reference appendices. The Epilogue reflects on what the 2026 Cursor reveals about software's near future, and what workflow discipline as a durable skill actually means in a field moving this fast.

Epilogue: From Editor with AI to Agent Runtime with an Editor

When Cursor launched in March 2023, it was a VS Code fork with a chat sidebar. Three years later, it's a development environment whose primary metaphor is no longer "an editor that helps you write code" but "an agent runtime that happens to have an editor surface." The shift is subtle in any single feature change and stark in the cumulative arc.

The 2023 Cursor: AI suggests text. You type. The model is a guest in your editor.

The 2024 Cursor: AI proposes multi-file changes. You review diffs. The model is a colleague who occasionally types for you.

The 2025 Cursor: AI runs autonomously in foreground and background. You direct, scope, and review. The model is a worker you delegate to.

The 2026 Cursor: AI runs in parallel. AI fixes its own previous outputs. AI orchestrates external systems. You design the workflows and audit the results. The model is the runtime.

That progression isn't unique to Cursor. The same arc shows up in Claude Code, in Copilot, in Windsurf, across the broader category. Cursor is the most-adopted instance of the category in 2026; the category itself is what's reshaping how software gets written.

What This Reveals About Software's Near Future

A few observations the 2026 Cursor makes hard to ignore:

The cost of writing code is dropping toward zero, and the cost of deciding what to write is staying flat. The leverage from AI editors compounds at writing time. The leverage at design, architecture, and verification time is much smaller. Software economics have always favored the rare ability to decide what to build over the common ability to type it; that asymmetry is intensifying.

Verification becomes the bottleneck. When the model can produce ten times the code per engineer-hour, engineer time shifts from typing to verifying. The skills that matter shift accordingly. Reading carefully, designing systems for verifiability, building strong test suites — these are the durable engineering skills of the next era.

Workflow discipline is the durable engineering skill. Specific tools change. Cursor 2027 will look different from Cursor 2026 in ways we can't fully predict. The discipline of scoping work tightly, picking the right mode for the right task, reading every line, maintaining shared rules — that survives across tool transitions. Engineers who internalize the discipline ride waves; engineers who don't fight each tool from scratch.

The editor is becoming infrastructure. Cursor at the team level isn't a "preferred editor"; it's a piece of shared infrastructure with team-wide configuration, shared rules, audited behavior. The transition from "what editor do you like" to "what editor does the team use" is mostly complete in segments where AI editors have penetrated. The category is converging on shared tooling, not personal preference.

The integration surface is getting larger. MCP, BugBot, Background Agents, Multitask — none of these existed three years ago. Each year of progress adds to the surface area where AI plugs into developer workflows. The Cursor of five years from now will probably look like a different product than today's Cursor — not just better at the same things, but different in shape.

Workflow Discipline as the Durable Skill

The book has come back to this idea repeatedly because it keeps being right. The engineers who get the most out of AI editors aren't the ones with the cleverest prompts or the deepest knowledge of every feature. They're the ones who:

- Scope their work tightly before invoking AI
- Pick the right mode for each task
- Read every diff carefully
- Maintain shared rules that anchor model behavior
- Recognize their own failure modes and the model's failure modes
- Ship code they understand, regardless of who typed it

These habits aren't specific to Cursor. They aren't specific to AI editors at all. They're the habits of engineers who care about the quality of what they ship — applied in a context where typing speed is much higher than it used to be.

If you take only one thing from this book, take that. Product features will rotate. Pricing will change. The model picker will look different in two years. The discipline endures.

A Closing Note

I wrote this book because I think Cursor is one of the most important tools to learn well in 2026, and because almost everyone using it is using it badly. The "badly" isn't an indictment — the product moves too fast for documentation to keep up, community advice is fragmented, and the people who use it best haven't always written down what they know.

The book is my attempt to write down what I've learned across the project of building, shipping, and reviewing real production code with this tool over the last few years. It's opinionated where opinions are useful and honest where honesty is necessary. It will be wrong in places by the time you read it; the field moves faster than book publishing.

But the workflow discipline endures. The "every line, always" habit endures. The mode rubric endures. The rules-driven approach endures. Internalize these and stay calibrated to the tool's continuing evolution, and you'll be ahead of most engineers using Cursor today and most engineers who'll be using whatever Cursor becomes tomorrow.

Turn the page. The reference material follows — keyboard shortcuts, a starter rules library, troubleshooting, glossary, the 2026 timeline, resources. Use it as you need it.

Good luck. Happy shipping.

308

Part IX: Reference

"The right reference is the one you can find when you need it."

— Anonymous

Appendix A: Keyboard Shortcuts and Interface Tips

This appendix collects the keyboard shortcuts most worth knowing and a handful of interface tips that don't fit cleanly elsewhere. Shortcuts are listed with macOS bindings first, then Windows/Linux. Where they match VS Code exactly, the VS Code binding is noted.

Shortcuts can be remapped in Settings → Keyboard Shortcuts. The defaults below assume an unmodified install.

AI Primitives

Action	macOS	Windows/Linux
Open chat sidebar	Cmd+L	Ctrl+L
Open inline edit (Cmd+K)	Cmd+K	Ctrl+K
Open Manual Edit	Cmd+I	Ctrl+I
Open Agent (full panel)	Cmd+Shift+I	Ctrl+Shift+I
Accept inline suggestion (Tab)	Tab	Tab
Reject inline suggestion	Esc	Esc
Accept Cmd+K diff	Cmd+Enter	Ctrl+Enter
Reject Cmd+K diff	Esc	Esc
New chat session	Cmd+N (in chat)	Ctrl+N
Switch chat mode	Cmd+.	Ctrl+.

Navigation

Action	macOS	Windows/Linux
Command palette	Cmd+Shift+P	Ctrl+Shift+P
Quick file open	Cmd+P	Ctrl+P
Go to symbol (file)	Cmd+Shift+O	Ctrl+Shift+O
Go to symbol (workspace)	Cmd+T	Ctrl+T
Go to definition	F12	F12

Action	macOS	Windows/Linux
Find all references	Shift+F12	Shift+F12
Toggle terminal	Cmd+`	Ctrl+`
Toggle sidebar	Cmd+B	Ctrl+B
Toggle chat sidebar	Cmd+L (toggle)	Ctrl+L

Editing

Action	macOS	Windows/Linux
Multi-cursor at click	Option+Click	Alt+Click
Add cursor above/below	Cmd+Option+Up/Down	Ctrl+Alt+Up/Down
Select next match	Cmd+D	Ctrl+D
Select all matches	Cmd+Shift+L	Ctrl+Shift+L
Move line up/down	Option+Up/Down	Alt+Up/Down
Duplicate line	Shift+Option+Up/Down	Shift+Alt+Up/Down
Comment toggle	Cmd+/	Ctrl+/
Format document	Shift+Option+F	Shift+Alt+F
Rename symbol (LSP)	F2	F2
Quick fix	Cmd+.	Ctrl+.

Selections

Action	macOS	Windows/Linux
Expand selection	Ctrl+Shift+Cmd+Right	Shift+Alt+Right
Shrink selection	Ctrl+Shift+Cmd+Left	Shift+Alt+Left
Select line	Cmd+L (in editor)	Ctrl+L
Select to end of line	Shift+Cmd+Right	Shift+End
Select to start of line	Shift+Cmd+Left	Shift+Home

Files and Tabs

Action	macOS	Windows/Linux
New file	Cmd+N	Ctrl+N
Save	Cmd+S	Ctrl+S

Action	macOS	Windows/Linux
Save all	Cmd+Option+S	Ctrl+K, S
Close tab	Cmd+W	Ctrl+W
Reopen closed tab	Cmd+Shift+T	Ctrl+Shift+T
Switch tabs	Cmd+1, Cmd+2...	Ctrl+1, Ctrl+2...
Cycle tabs	Cmd+Option+Right/Left	Ctrl+Tab
Split editor	Cmd+\	Ctrl+\

Search

Action	macOS	Windows/Linux
Find in file	Cmd+F	Ctrl+F
Replace in file	Cmd+Option+F	Ctrl+H
Find in workspace	Cmd+Shift+F	Ctrl+Shift+F
Replace in workspace	Cmd+Shift+H	Ctrl+Shift+H

Diff Review

Action	macOS	Windows/Linux
Accept hunk	Cmd+Enter	Ctrl+Enter
Reject hunk	Cmd+Backspace	Ctrl+Backspace
Next hunk	F7	F7
Previous hunk	Shift+F7	Shift+F7
Accept all in file	Cmd+Option+Enter	Ctrl+Alt+Enter
Reject all in file	Cmd+Option+Backspace	Ctrl+Alt+Backspace

Mode-Switching Tips

When chat is open, the mode dropdown at the bottom of the panel lets you switch among Ask, Manual, and Agent. Cmd+. (Ctrl+.) cycles through modes from anywhere in the chat input.

The mode persists per-session, not globally. A new chat starts in the last-used mode by default.

For inline use, Cmd+K is always Cmd+K regardless of the chat mode — the inline-edit primitive is independent of the chat panel.

Power-User Navigation

A few patterns that pay back over time:

Cmd+T (Go to Symbol Workspace) is faster than typing `@symbol` in chat. When you want to find a function across the project, the workspace symbol picker is sub-second; chat takes longer.

Multi-cursor with Cmd+D (select next match) is faster than asking the model to rename. For mechanical multi-cursor edits the IDE can do itself, skip the AI round-trip.

Quick file open (Cmd+P) supports glob-like prefixes. "@" goes to symbols, "#" goes to lines, ":" goes to a specific column. Combine: `@searchProducts:42` jumps to line 42 of the searchProducts symbol.

Split editors for diff comparison. Cmd+\ splits horizontally. Useful when reviewing large Manual diffs alongside the source file.

Cmd+B (toggle sidebar) for screen real estate. The file tree disappears; the editor expands. Useful in deep-focus work where you don't need navigation visible.

Customization

A few customizations worth considering after you've used Cursor for a few weeks:

Rebind the chat shortcut if it conflicts. Cmd+L is also "go to line" in some VS Code workflows. If you've remapped, the chat shortcut has to move.

Add a shortcut for "switch chat to Agent." Some users add a single shortcut that opens chat in Agent mode directly, skipping the mode-switch step. Useful if you reach for Agent often.

Configure per-language Tab settings if Tab is too aggressive in some contexts. Settings → Tab Autocomplete → Language Settings.

Turn off auto-save if it conflicts with hooks. If you have post-save hooks that you want firing only on explicit saves, disable auto-save in the Cursor-specific settings.

Interface Tips

A handful of small things that matter:

The bottom-right indexing indicator. Watch it on a fresh project — when it changes from "indexing..." to a checkmark, the index is ready. Don't judge AI quality before then.

The credit indicator in the chat panel. Shows how much of your monthly pool you've used. Glance at it once a day.

Mode-specific colored borders on chat. Ask is blue, Manual is amber, Agent is red (the colors evolve, but the principle holds — modes are visually distinct). Don't accidentally Agent into a task that should have been Ask.

The stop button on running Agent or Background Agent. Bigger than it looks; you can't accidentally stop a run, but it's there when you need it.

The "@" menu in chat input. Comprehensive — files, folders, symbols, web, docs, MCP servers. Browse it once to see what's available.

The diff review tree. Per-file, per-hunk granularity. Don't accept or reject the whole diff at once unless you've checked each part.

Appendix B: Project Rules Library

This appendix is a 40-plus entry starter library of `.cursor/rules/` files for common stacks. Use them as starting points; expect to refine for your project's specific conventions. Each rule is shown as the file's content; save under the suggested filename in `.cursor/rules/` at your project root.

The rules use the `.mdc` format with frontmatter. Adapt the `globs` field to match your project's directory structure.

TypeScript / React

`typescript-general.mdc`

```
---
description: General TypeScript conventions
globs:
  - "**/*.ts"
  - "**/*.tsx"
alwaysApply: true
---

# TypeScript Conventions

## Imports
- Use named imports, not default imports, for project code.
- Import order: external libraries, then internal `@/` paths, then relative.
- No unused imports.

## Types
- Prefer `type` aliases over `interface` for object shapes.
- All exported functions have explicit return type annotations.
- Avoid `any`. Use `unknown` and narrow with type guards.
- Avoid type assertions (`as Foo`) except at boundaries with external data.

## Naming
- Variables and functions: camelCase.
- Types and components: PascalCase.
- Constants: SCREAMING_SNAKE_CASE only for true compile-time constants.
- Boolean variables and props: prefix with `is`, `has`, `should`.
```

```
## Async
- Always `await` Promises. Never leave them dangling.
- Prefer `async/await` over `.then()` chains.
- Wrap top-level awaits in IIFE if needed for module-level setup.

## Errors
- Throw `Error` instances, not strings. Custom error classes for domain errors.
- Don't catch errors you can't handle. Let them propagate to error boundaries.
```

react-components.mdc

```
---
description: React component conventions
globs:
  - "**/*.tsx"
  - "**/components/**/*.ts"
alwaysApply: false
---

# React Components

## Component shape
- Function components only. No class components.
- Single default export per file (the component); other exports are types or helpers.
- Props typed via a `Props` type alias defined in the same file.

## Hooks
- Custom hooks named `useXxx`.
- Don't call hooks conditionally.
- Memoize expensive computations with `useMemo`; memoize callbacks with `useCallback` only when the callback identity matters.

## State
- Local state via `useState` for simple values, `useReducer` for complex.
- Server state via TanStack Query, not local state.
- Don't lift state higher than necessary.

## JSX
- Self-closing tags for elements without children.
- Conditional rendering via `&&` for show/hide; ternary for either-or.
- Avoid inline functions in props when memoization matters.

## Styling
```

```
- Tailwind classes inline. Don't extract to className constants unless reused.
- For complex conditional classes, use the project's `cn()` helper.
```

react-testing.mdc

```
---
description: React testing conventions with React Testing Library and Vitest
globs:
  - "**/*.test.tsx"
  - "**/*.spec.tsx"
alwaysApply: false
---

# React Testing

## Framework
- Vitest, not Jest. Use `vi` for mocks.
- React Testing Library for rendering and queries.
- `@testing-library/user-event` for interactions, not `fireEvent`.

## Test names
- "should [behavior] when [condition]" format.
- Group related tests in `describe` blocks named after the component.

## Queries
- Prefer `getByRole` over `getByText`.
- Use `findBy` for async-appearing elements.
- `queryBy` only when asserting absence.

## Mocks
- Mock external services with `@/test/mocks`. Never call real APIs in unit tests.
- For network requests, use MSW handlers in `tests/handlers/`.
```

Python / FastAPI

python-general.mdc

```
---
description: General Python conventions
globs:
  - "**/*.py"
alwaysApply: true
---

# Python Conventions
```

```
## Style
- Follow PEP 8. Line length 100.
- Type hints on all public functions and methods.
- f-strings for string formatting. No `.format()` or `%` interpolation.

## Imports
- Group: stdlib, third-party, project. Blank line between groups.
- No wildcard imports.
- No relative imports beyond a single `.` (use absolute from project root).

## Functions
- Pure functions where possible. Side effects isolated to specific layers.
- Default to `Optional[T]` over `T | None` for older codebase compatibility.
- Mutable default arguments are forbidden. Use `None` and check.

## Errors
- Custom exceptions in `errors.py` per package, inherit from a project base class.
- Don't catch `Exception` broadly. Catch specific types.
- Re-raise with `raise NewError(...) from original_error` to preserve chain.
```

fastapi.mdc

```
---
description: FastAPI patterns
globs:
  - "api/**/*.py"
  - "src/api/**/*.py"
alwaysApply: false
---

# FastAPI

## Routes
- Routes return Pydantic models, not dicts.
- Path parameters in plural for collections, singular for items.
- Use dependency injection (`Depends()`) for auth, db sessions, settings.

## Response models
- Always specify `response_model` in the route decorator.
- Don't return ORM objects directly; convert to Pydantic models first.

## Errors
- Raise `HTTPException` for HTTP errors. Custom domain exceptions get mapped via
```

```
 exception handlers.
- 404 for not found, 422 for validation, 401 for auth, 403 for permission.

## Validation
- Use Pydantic v2 syntax (`model_validate`, `model_dump`).
- Validators on Pydantic models, not in route handlers.
```

Go / gRPC

go-general.mdc

```
---
description: General Go conventions
globs:
  - "**/*.go"
alwaysApply: true
---

# Go Conventions

## Style
- gofmt + goimports run automatically.
- Comments on exported identifiers start with the name.
- Errors returned as the last return value.

## Errors
- Wrap with `fmt.Errorf("doing X: %w", err)` to preserve chains.
- Compare with `errors.Is` and `errors.As`, not `==` or type assertions.
- Don't ignore errors. Either handle, log, or propagate.

## Concurrency
- Don't share memory without synchronization.
- Channels for ownership transfer; mutexes for shared state.
- Always have a way to stop goroutines (context, done channel).

## Naming
- Short, descriptive names. `i` for index, `err` for error, `ctx` for context.
- Acronyms uppercased: `userID`, `httpClient`, `apiKey`.
```

grpc.mdc

```
---
description: gRPC service conventions
globs:
  - "**/proto/**"
```

```
  - "**/*_grpc.go"
alwaysApply: false
---

# gRPC

## Proto definitions
- Each service in its own `.proto` file.
- Field numbering preserves backward compat. Never renumber.
- New fields are optional; required-by-convention via service contract.

## Service implementation
- Method handlers small; delegate to package functions.
- Always pass `context.Context` through.
- Return appropriate `codes.X` for errors, not `errors.New(...)`.

## Streaming
- Send-side: `Send` returns; check error.
- Recv-side: `Recv` returns; handle EOF as normal close.
```

Rust / Axum

rust-general.mdc

```
---
description: General Rust conventions
globs:
  - "**/*.rs"
alwaysApply: true
---

# Rust Conventions

## Style
- rustfmt run automatically.
- Clippy with `--all-targets` clean before commit.

## Errors
- `thiserror` for library code. `anyhow` for application code.
- `Result<T, MyError>` over `Result<T, Box<dyn Error>>` for typed errors.
- Don't `unwrap` in production code. Use `?` and handle properly.

## Ownership
- Prefer borrowing over cloning. Clone only when needed.
```

```
- `&str` over `String` in function args unless ownership is required.
- Lifetimes elided where possible; explicit when ambiguous.

## Async
- `tokio` runtime. `async fn` and `.await` throughout.
- Don't block in async contexts. Use `spawn_blocking` for sync work.
```

Mobile

swift-ios.mdc

```
---
description: Swift / iOS conventions
globs:
  - "**/*.swift"
alwaysApply: false
---

# Swift / iOS

## Style
  SwiftLint clean.
- Trailing closure syntax for last closure arg.
- Implicit returns in single-expression closures and computed properties.

## Concurrency
  async/await for new code; Combine for reactive UI flow.
- `@MainActor` annotations for UI-touching code.
- No completion handlers in new code (only legacy bridges).

## SwiftUI
- View extensions over view modifiers for reuse.
- `@StateObject` for owned state; `@ObservedObject` for injected.
- Don't put expensive logic in `body`.
```

kotlin-android.mdc

```
---
description: Kotlin / Android conventions
globs:
  - "**/*.kt"
alwaysApply: false
---

# Kotlin / Android
```

```
## Style
- ktlint clean.
- Data classes for value objects.
- Sealed classes for closed type hierarchies.

## Concurrency
- Coroutines, not RxJava (for new code).
- `Flow` for streams; `StateFlow` for state.
- Scope coroutines properly: `viewModelScope`, `lifecycleScope`.

## Compose
- Stateless composables. State hoisted to ViewModel.
- `remember` for view-local cache; `rememberSaveable` for surviving config chang
es.
- Use Modifier from parameters; don't construct internally.
```

Infrastructure

terraform.mdc

```
---
description: Terraform conventions
globs:
  - "**/*.tf"
  - "**/*.tfvars"
alwaysApply: false
---

# Terraform

## Module structure
- One module per directory. README.md documents inputs and outputs.
- Variables in `variables.tf`. Outputs in `outputs.tf`. Resources in `main.tf`.

## State
- Remote state backend (S3 + DynamoDB) for shared state.
- Don't commit `.tfstate` files. Don't commit `.terraform/` directory.

## Resources
- All resources tagged with `Environment`, `Owner`, `CostCenter`.
- Use `lifecycle` blocks to prevent accidental destruction of stateful resources.
- `count` and `for_each` carefully; don't shuffle resources between branches.
```

kubernetes.mdc

```
---
description: Kubernetes manifest conventions
globs:
  - "k8s/**/*.yaml"
  - "deploy/**/*.yaml"
alwaysApply: false
---

# Kubernetes

## Manifests
- One resource per file (Deployment, Service, ConfigMap separate).
- Labels: app, component, version, environment.
- Resource limits and requests on every container.

## Secrets
- Don't commit secrets in plaintext. Reference SealedSecrets or external secrets.
- ConfigMaps for non-sensitive config; Secrets for credentials.

## Health
- Readiness and liveness probes on every Deployment.
- Probe paths defined per service.
```

Test Discipline

tests-general.mdc

```
---
description: Test conventions across languages
globs:
  - "**/*.test.*"
  - "**/*.spec.*"
  - "**/test_*.py"
  - "**/*_test.go"
alwaysApply: true
---

# Test Discipline

## Coverage
- Every new behavior has a test.
- Tests cover happy path, edge cases, and error paths.
```

```
- Don't write tests just for coverage. Tests should verify meaningful behavior.

## Names
- Test names describe behavior, not implementation.
- "should [behavior] when [condition]" or "test_[behavior]_when_[condition]".

## Setup
- No real network calls in unit tests. Mock at boundaries.
- Test data factories for complex setup. No magic literals scattered through tests.
- Each test cleans up after itself; no test depends on another's state.

## Assertions
- One logical assertion per test. Multiple `expect(...)` are fine if they're aspects of the same behavior.
- Assert specific values, not "anything truthy."
```

Security

security.mdc

```
---
description: Security baseline rules
globs:
  - "**/auth/**"
  - "**/security/**"
  - "**/payment/**"
  - "**/admin/**"
alwaysApply: false
---

# Security Baselines

## Authentication
- All routes default to authenticated. Public routes are explicit opt-out.
- Sessions expire on idle. Refresh tokens rotate on use.
- No auth state in client-side localStorage.

## Authorization
- Authorization checks at every protected resource access.
- Don't trust the client's claim of role; verify server-side.
- Audit log every privileged action.

## Input
```

```
- Validate all user input at the API boundary.
- Parameterize all database queries. Never concatenate user input into SQL.
- Sanitize output destined for HTML or JSON contexts.

## Secrets
- No secrets in code. Reference environment variables or secret manager.
- No secrets in logs. Use redaction.
- Rotate any secret that has been logged or committed.
```

Git Conventions

git-conventions.mdc

```
---
description: Git workflow conventions
globs: []
alwaysApply: true
---

# Git Conventions

## Commits
- Conventional commits format: `type(scope): description`.
- Types: feat, fix, refactor, docs, test, chore, style.
- Scope optional; description in present tense, no trailing period.

## Branches
- `feat/<description>`, `fix/<description>`, `refactor/<description>`.
- Slash separator. Lowercase. Hyphens for multi-word.

## PRs
- Title matches the conventional commit format.
- Description includes: motivation, approach, testing notes.
- Link relevant issues.
- Squash merge by default.
```

Documentation

docs.mdc

```
---
description: Documentation conventions
globs:
  - "**/*.md"
```

```
  - "docs/**"
alwaysApply: false
---

# Documentation

## Voice
- Second person ("you do X"), not first ("I do X") or third ("the user does X").
- Direct and short. Cut adjectives that aren't load-bearing.
- Headings as descriptions of content, not commands.

## Structure
- Each doc has a one-paragraph summary at the top.
- Code blocks have language tags for syntax highlighting.
- External links: full URLs, no shorteners.
```

How to Use This Library

The starter library captures common conventions; your project's conventions are specific. The workflow that produces useful per-project rules:

1. Drop the relevant starters into `.cursor/rules/`.
2. Use the project for a week.
3. Notice where the model is producing output that doesn't match your conventions.
4. For each pattern, either: (a) update the rule to capture your actual convention, or (b) loosen the rule if it's stricter than reality, or (c) remove the rule if it doesn't apply.
5. Repeat monthly.

After a few iterations, the rules are tailored to your project. The model's output starts feeling like it was written by someone on your team, because the rules anchor it there.

Appendix C: Troubleshooting Common Issues

This appendix covers the issues that come up most often in real Cursor use, organized by symptom. For each, the likely causes and the steps to resolve. When something is breaking, work through the steps in order — most issues resolve at the first or second step.

Indexing Stuck or Slow

Symptom: The indexing indicator has been spinning for ten minutes on a small project, or for an hour on a large one without visible progress.

Likely causes: 1. The codebase is genuinely large and indexing is in progress (just slow). 2. `.cursorignore` is missing or incomplete; Cursor is trying to index node_modules, build outputs, generated code. 3. An antivirus or filesystem watcher is interfering. 4. A single very-large file (5,000+ lines) is slowing the pipeline.

Resolution steps:

1. Open Settings → Codebase → Indexing. Check the status. Errors? Specific files stuck?
2. Check `.cursorignore`. If absent, create it with at least: `node_modules/`, `dist/`, `build/`, `.next/`, `*.lock`. Commit and reload.
3. Force a fresh re-index: Settings → Codebase → Re-index Project.
4. If specific files are stuck, exclude them in `.cursorignore`. Re-index.
5. If the project is genuinely huge, scope indexing to subdirectories: Settings → Codebase → Scoped Paths.
6. Check antivirus exclusions. Add the project directory and `~/.cursor/` (or the Windows/Linux equivalent) to AV exclusions.
7. Try the project on a different filesystem. Network filesystems and encrypted drives slow indexing substantially.

Tab Autocomplete Is Weak or Missing

Symptom: Tab proposes generic completions, or none at all, in code where it should be productive.

Likely causes: 1. Indexing isn't complete yet. 2. The file is excluded from indexing by `.cursorignore`. 3. The file is too large (over 5,000 lines). 4. Network or service issue with Cursor's Tab model. 5. Tab is configured off for this file type.

Resolution steps:

1. Check the indexing indicator. If still spinning, wait.
2. Confirm the file isn't in `.cursorignore`. If unintentionally excluded, fix it.
3. Check the file size. Over 5,000 lines means weaker Tab; consider splitting.
4. Check Cursor's status page (cursor.com/status) for service issues.
5. Check Settings → Tab Autocomplete → file type filters. Make sure your language is enabled.
6. Reload the editor (Cmd+Shift+P → "Reload Window"). Sometimes the connection state needs a reset.

Agent Calls Timing Out

Symptom: Agent runs hang after a few minutes, or report a timeout error.

Likely causes: 1. The task is genuinely large and the model is taking its time. 2. Network instability on your side. 3. Cursor service degradation. 4. Agent is stuck in a loop and not making progress. 5. A command Agent is running has hung.

Resolution steps:

1. Check the chat log for what Agent was doing when it stopped. If it's stuck on a specific command, that command may have hung.
2. Pause and resume the run. Sometimes that unsticks transient issues.
3. Check Cursor's status page.
4. Check your network. A flaky connection during long Agent runs causes timeouts.
5. If Agent is in a loop (re-trying the same fix), stop the run and restart with tighter scope. The loop won't resolve itself.

6. Try a different model. Sometimes one model gets stuck on a task another handles cleanly.

Credit Pool Exhaustion

Symptom: "You've used your monthly credits" notification, or AI features stop working mid-month.

Likely causes: 1. Heavy use without picker discipline (Opus on everything, Agent for everything). 2. A runaway Agent run consumed substantial credits. 3. Multiple Multitask runs without budget awareness. 4. Hitting a hard cap you set (intentionally or forgot about).

Resolution steps:

1. Check the usage dashboard. Identify what consumed the credits – model, mode, frequency.
2. If you have pay-as-you-go enabled, you can continue at standard rates. Decide whether to enable it.
3. If you have a hard cap, either raise it or accept reduced usage for the rest of the month.
4. Adjust workflow: switch default model to DeepSeek V4 Pro, default mode to Manual instead of Agent, lean harder on Tab and Cmd+K.
5. Consider upgrading to Pro+ or Ultra if your usage consistently exceeds the included credits.
6. For team plans, the admin sees the breakdown and can pool-balance across users.

MCP Install Failures

Symptom: An MCP server fails to install, or installs but doesn't show up in the tool list.

Likely causes: 1. The MCP server's runtime requirements aren't met (Node version, Python version). 2. Authentication failed during the OAuth flow. 3. A custom server's configuration in `mcp.json` has a typo. 4. Network issue reaching the server's package source.

Resolution steps:

1. For catalog installs: re-run the install flow. The OAuth step sometimes needs a retry.
2. For custom servers: check `.cursor/mcp.json` for syntax errors. Validate as JSON.
3. Check the MCP server's prerequisites. Many require specific Node or Python versions.
4. Look at Cursor's MCP logs: Command Palette → "Show MCP Logs." Errors usually surface there.
5. For stdio-based servers, run the server's command manually in a terminal. If it doesn't run there, it won't run from Cursor.
6. For HTTP-based servers, confirm the URL is reachable from your network.

Background Agent Stuck States

Symptom: A Background Agent run has been "running" for hours past expected completion, or shows a state that doesn't match what's happening.

Likely causes: 1. The Agent genuinely is taking that long (rare for runs over 1 hour). 2. The sandbox infrastructure had an issue and the run is stuck. 3. The Agent is waiting for input it never received. 4. A network issue between Cursor and the GitHub integration is blocking PR creation.

Resolution steps:

1. Open the Background Agent dashboard. Look at the run's log. What was the last activity?
2. If the log shows recent activity, the run is probably making progress. Wait.
3. If the log has been silent for 30+ minutes, the run is probably stuck. Cancel it.
4. Check whether a partial PR was created. Sometimes Background Agent creates the branch but fails on the PR step. Open the PR manually if so.
5. Open a support ticket if the run state is permanently stuck and your work is blocked.

Login and Sync Issues

Symptom: Cursor reports you're logged out, or your settings/extensions aren't syncing correctly.

Likely causes: 1. Session expired and needs re-auth. 2. SSO token expired (for Teams/Enterprise). 3. Multiple Cursor installs (e.g., dev and stable channels) competing for the same session. 4. Local cache corruption.

Resolution steps:

1. Sign out fully. Sign back in.
2. For SSO users: check that your IdP session is current. Re-authenticate via your SSO provider.
3. If you have multiple Cursor installs, sign out of all of them and sign back into just the one you want.
4. Clear Cursor's local cache: Command Palette → "Cursor: Clear Cache." Restart.
5. Check that your account is in good standing — billing failures can produce logout-like states.

Privacy Mode Behaving Unexpectedly

Symptom: You enabled Privacy mode but data still seems retained, or features aren't working under Privacy mode.

Likely causes: 1. The setting was enabled but only for new requests. Old data is still in the pipeline. 2. Some features have known limitations under Privacy mode. 3. Org-level enforcement overrides the per-user setting (Enterprise plans). 4. The feature you're using bypasses Privacy mode (rare; usually documented).

Resolution steps:

1. Verify the setting is on: Settings → Privacy → Privacy Mode. Should show as enabled.
2. Wait for the pipeline to clear. New requests under Privacy mode are zero-retention; previously-stored data is on its retention schedule.
3. For Enterprise users: check the org-level policy. Your admin may have specific Privacy mode settings.

4. Read the feature's documentation. Some experimental or beta features may not yet support Privacy mode.

File Save Doesn't Apply Changes

Symptom: You hit save, but the file doesn't update on disk, or the next AI invocation doesn't see your changes.

Likely causes: 1. The file was modified externally and Cursor's view is stale. 2. A hook is intercepting and modifying the saved content. 3. Auto-format on save is rewriting in unexpected ways. 4. A file watcher race condition.

Resolution steps:

1. Close and reopen the file. Cursor should re-read from disk.
2. Check the hooks directory. A misbehaving hook can corrupt the save.
3. Disable format-on-save temporarily to isolate the issue.
4. Force a re-index of the file: right-click → "Re-index File."
5. Check the file's timestamp on disk. If it doesn't match the save, the save isn't reaching disk.

Cursor Won't Start

Symptom: Launching Cursor produces an error, hangs on the splash screen, or crashes immediately.

Likely causes: 1. A corrupt installation. 2. An incompatible extension. 3. A corrupted workspace state. 4. OS-level permissions issue.

Resolution steps:

1. Try opening Cursor without any project: launch from a fresh location and don't open a recent project.
2. Disable extensions: launch with the "Cursor: Start with Extensions Disabled" option (varies by platform).
3. Clear Cursor's user state: rename `~/.cursor/` (or the platform equivalent) and let Cursor recreate it. Re-sign in.
4. Reinstall Cursor.

5. Check the OS logs for crash details. macOS: Console.app. Windows: Event Viewer. Linux: journalctl.
6. Open a support ticket with the crash details if all else fails.

Performance Is Slow Across the Board

Symptom: Cursor feels sluggish — typing latency, slow file open, slow search, slow autocomplete.

Likely causes: 1. Too many extensions consuming resources. 2. Indexing is in progress and consuming CPU. 3. Large files open in the editor (many tabs). 4. Antivirus scanning slowing every file operation. 5. Low-memory pressure on the machine.

Resolution steps:

1. Check Activity Monitor/Task Manager/`top`. What's Cursor consuming? RAM? CPU?
2. Disable extensions one by one to find the culprit.
3. Close tabs you don't need open. Dozens of accumulated tabs have real cost.
4. Check antivirus exclusions for the project directory.
5. Restart Cursor to clear accumulated state.
6. Check available RAM. Cursor on a project with a million files needs at least 8GB free; ideally 16GB.

When to Open a Support Ticket

If the steps above don't resolve the issue, open a support ticket through Cursor. Include:

- The version of Cursor (Help → About).
- Your operating system and version.
- The plan you're on.
- A clear description of the symptom.
- Steps to reproduce, if reproducible.
- Logs from the relevant Cursor logs panel.

For Enterprise users, your CSM is the faster path. For Pro and Teams users, the in-app support flow reaches the support team within a business day.

Appendix D: Glossary of Terms

The vocabulary that comes up across this book and across Cursor's broader ecosystem. Definitions kept short and practical.

Agent (mode). The autonomous chat mode where the model reads files, edits files, runs terminal commands, and iterates through a multi-step task without per-step user input. See Chapter 8.

AI-native IDE. A development environment built around AI primitives as the primary interaction surface, not as bolted-on extensions. Cursor, Windsurf, and a few competitors fit. See Chapter 1.

AI-Code-Tracking API. Cursor Enterprise's API for identifying and reporting which code in a repository was AI-generated, with metadata about which engineer, model, and mode produced it. See Chapter 28.

Anysphere. The company that makes Cursor. Founded in 2023 by four MIT graduates.

Ask (mode). The read-only chat mode. The model can read your codebase and answer questions but cannot edit files or run commands. See Chapter 6.

Auto (model). The "let Cursor choose" option in the model picker. Picks per-task based on heuristics. See Chapter 18.

Background Agent. A cloud-hosted Agent run that executes asynchronously on Cursor's infrastructure rather than locally in the editor. You kick off, walk away, return to a finished PR. See Chapter 16.

BugBot. Cursor's automated PR review and fix tool. Reads PRs, identifies issues, opens fix PRs (since the February 2026 graduation from beta). Priced separately from the editor subscription. See Chapter 23.

BYOK (Bring Your Own Key/API). The configuration that routes Cursor's model invocations through your own provider account (Anthropic, OpenAI, self-hosted) rather than Cursor's credit pool. See Chapter 20.

Catalog (MCP). Cursor's curated registry of pre-vetted MCP servers with one-click installation. See Chapter 22.

Claude Code. Anthropic's terminal-native coding agent. A different product from Cursor (covered in *Claude Code: The Power User Guide*). See Chapter 2.

Cline. An open-source VS Code extension implementing Agent-mode-style features with BYOK pricing. A Cursor alternative. See Chapter 2.

Cmd+K (or Ctrl+K). The inline-edit shortcut. Opens an inline prompt at your cursor or selection. See Chapter 11.

Composer. The original name of Manual Edit mode (2024). Still seen in older blog posts and tutorials. See Chapter 7.

Context retrieval. The process of pulling relevant code chunks from the codebase index into the model's context for a request. The mechanism that makes Cursor codebase-aware. See Chapter 13.

Credit pool. The monthly dollar-denominated allocation of model usage in Cursor's pricing model. Replaced fixed request quotas in mid-2025. See Chapter 19.

Credits. The unit of usage in Cursor's pricing. Roughly equivalent to dollars at standard provider rates. See Chapter 19.

`.cursor/`. The per-project directory at your repo root containing rules, ignore configurations, hooks, slash commands, Skills, and MCP configurations. Checked into version control. See multiple chapters.

`.cursorignore`. The file that controls which paths are excluded from Cursor's codebase index. Syntax compatible with `.gitignore`. See Chapter 15.

`.cursorrules`. The legacy single-file rules format, predecessor to `.cursor/rules/`. Still supported but new projects use the directory-based format. See Chapter 14.

DeepSeek V4 Pro. The DeepSeek frontier model in Cursor's picker as of 2026. The cost-quality leader for most tasks; roughly one-tenth the cost of Claude Opus on coding workloads. See Chapter 18.

Diff review. The interface where Manual or Agent's proposed changes are presented for accept/reject decisions. Per-file, per-hunk granularity. See Chapter 12.

Embedding. A vector representation of a file or code chunk's semantic content, used for similarity-based retrieval in the codebase index. See Chapter 13.

Enterprise (plan). Cursor's top tier, for organizations needing SCIM, audit logs, custom contractual terms, and similar enterprise-grade capabilities. Negotiated pricing. See Chapter 28.

Fast model. Cursor's tuned model used for Tab autocomplete and some default Cmd+K actions. Free at the per-invocation level on paid plans. See Chapter 18.

Fixer (BugBot). The capability added to BugBot in February 2026 — opening fix PRs to address issues BugBot identifies, not just commenting. See Chapter 23.

Foreground Agent. Agent mode running in your local editor, where you watch the loop and can intervene. Distinguished from Background Agent. See Chapter 8.

Gemini 3.1 Pro. Google's frontier model in Cursor's picker. The long-context champion (1M+ tokens). See Chapter 18.

GPT-5.4. OpenAI's current frontier in Cursor's picker. The "safe broad default" model. See Chapter 18.

Hard cap. A spending limit that, when reached, stops new credit consumption regardless of pay-as-you-go settings. The recommended safety mechanism for all paid plans. See Chapter 19.

Hobby (plan). Cursor's free tier. Functional for evaluation and very light use; insufficient for serious daily work. See Chapter 3.

Hooks. Shell scripts in `.cursor/hooks/` that fire automatically on specific Cursor events (pre-prompt, post-edit, etc.). See Chapter 24.

Index. Cursor's representation of your codebase — file tree, embeddings, symbol metadata, Merkle tree of file hashes. See Chapter 13.

Indexing. The process of building or updating the codebase index. See Chapter 13.

Manual Edit (mode). The chat mode where the model proposes coordinated multi-file changes for you to review and accept. Single-turn from the model's perspective. Renamed from Composer in 2025. See Chapter 7.

MCP (Model Context Protocol). The open protocol for AI clients to communicate with external context providers. The standard underlying Cursor's external integrations. See Chapter 22.

MCP server. A program implementing the MCP protocol, exposing tools and resources to AI clients. See Chapter 22.

Merkle tree. The hash-based data structure Cursor uses to detect which files have changed since the last index, enabling fast incremental re-indexing. See Chapter 13.

Mode switching. The practice of changing chat mode (Ask/Manual/Agent) mid-task to fit the current need. See Chapter 9.

Model picker. The dropdown for choosing which underlying frontier model handles a given chat or Cmd+K request. See Chapter 18.

`/multitask`. The Cursor 3.0 slash command that spawns parallel async subagents in isolated worktrees. See Chapter 17.

Open VSX. The open-source extension registry Cursor uses (instead of Microsoft's VS Code Marketplace, which has licensing restrictions). See Chapter 4.

Opus 4.7. Claude Opus 4.7, Anthropic's frontier reasoning model in Cursor's picker. Highest cost; strongest on architectural and complex reasoning tasks. See Chapter 18.

Pay-as-you-go. The setting that allows credit consumption beyond the included monthly amount, billed at standard rates. Optional; off by default. See Chapter 19.

Per-project rules. Rules in `.cursor/rules/` checked into the repo, applying only to that project. See Chapter 14.

Plaintext retention. The retention of unencrypted code by Cursor's infrastructure for purposes beyond the immediate request. Privacy mode prevents this. See Chapter 21.

Pool (credit). The team-level shared credit allocation in Teams and Enterprise plans, replacing per-user caps. See Chapter 27.

Privacy mode. The Cursor account setting for zero-data-retention infrastructure. The recommended default for working developers. See Chapter 21.

Pro (plan). Cursor's entry paid plan. $20/month with $20 of included credits. The right plan for most working developers. See Chapter 3.

Pro+ (plan). The middle paid tier. $60/month with 3x Pro's credit pool. For users who outgrow Pro. See Chapter 3.

Project rules. Configuration in `.cursor/rules/*.mdc` that shape model behavior for the project. See Chapter 14.

Read-act-observe loop. Agent mode's iteration cycle: read files/output, decide on an action, perform it, observe results, repeat. See Chapter 8.

SCIM (System for Cross-domain Identity Management). The protocol for automated user provisioning between IdPs and applications. Available on Enterprise plans. See Chapter 28.

Selection. Code highlighted in the editor before invoking Cmd+K, Manual, or Agent. Defines the editable scope. See Chapter 12.

Skill. A bundled package of rules, slash commands, and templates in `.cursor/skills/`. The unit of reusable workflow. See Chapter 24.

Slash command. A named prompt in `.cursor/commands/` invokable from chat with `/<name>`. See Chapter 24.

SSE (Server-Sent Events). One of three transports for MCP servers; an HTTP-based event stream. See Chapter 22.

Stdio (transport). One of three transports for MCP servers; the server runs as a child process communicating via stdin/stdout. See Chapter 22.

Streamable HTTP. The newer MCP transport (2026), production-grade for remote servers. See Chapter 22.

Sub-processor. A third-party service Cursor uses (model providers, hosting, etc.). Relevant for compliance review. See Chapter 31.

Tab. Cursor's inline autocomplete feature. Powered by Cursor's own tuned fast model. The most-used AI feature in Cursor. See Chapter 10.

Teams (plan). Cursor's small-to-medium team plan. $40/user/month with shared rules, pooled credits, SSO. See Chapter 27.

Turbopuffer. The vector database Cursor uses for storing codebase embeddings (as of 2026). See Chapter 13.

Ultra (plan). Cursor's highest individual plan. $200/month with 20x Pro's credit pool and priority access. See Chapter 3.

Verification (Agent). The commands Agent runs to confirm its work — typically tests, type-check, and lint, configured via `.cursor/environment.json`. See Chapter 16.

Windsurf. A direct Cursor competitor — also a VS Code fork with AI-native features. See Chapter 2.

Worktree. Git's mechanism for multiple working copies sharing the same `.git` directory. Used by `/multitask` to isolate parallel subagents. See Chapter 17.

Zero-data-retention. Infrastructure designed to process data without storing it beyond the request. Privacy mode's structural guarantee. See Chapter 21.

Appendix E: What's New in 2026

A timeline of significant Cursor updates through the period the book was written. The list covers the major ones, not exhaustively — Cursor ships patch-level updates frequently, and most don't make this list.

Cursor 3.0 (Q1 2026)

The headline release of early 2026, shipped in stages through January and February.

`/multitask` and async subagents. Parallel agent runs in isolated worktrees, with a multi-pane review interface for picking the winner. The most consequential workflow change since Background Agents. Detailed in Chapter 17.

MCP one-click installs from a curated catalog. The catalog vetted dozens of MCP servers (GitHub, Linear, Sentry, Postgres, Notion, Stripe, Slack, etc.) and made them installable without manual configuration. Detailed in Chapter 22.

Streamable HTTP MCP transport. A new transport better suited to production-grade remote MCP servers. Coexists with the older stdio and SSE transports.

Background Agent capacity expansions. Faster cold-starts, more parallel runs per user, longer maximum run durations.

Updated diff review interface. Clearer per-hunk controls, faster multi-file navigation, "ask follow-up about this hunk" surfacing more prominently.

BugBot Graduation to Fixer (February 2026)

BugBot moved from beta to general availability with the new "fixer" capability — opening its own fix PRs to address the issues it identifies, not just commenting. Reported 80%+ resolution rate on issues BugBot flags. Detailed in Chapter 23.

DeepSeek V4 Pro in the Picker (April 2026)

DeepSeek released V4 Pro on April 24, 2026, and Cursor added it to the model picker shortly after. Joins Claude Opus 4.7, GPT-5.4, and Gemini 3.1 Pro as the four frontier options. Significantly cheaper than the others, with comparable quality on most coding workloads. Detailed in Chapter 18.

DeepSeek V4 Flash arrived alongside it as the cheaper, faster variant.

Indexing Performance Improvements (Throughout 2026)

A series of incremental improvements to indexing performance, particularly aimed at monorepos:

- Parallel embedding generation reducing initial index time on large repos.
- Better incremental detection avoiding unnecessary re-indexing on branch switches.
- Improved summarization quality for very large files.
- Per-package scoping options for monorepos.

The cumulative effect is roughly a 2–3x speedup in initial indexing time on large monorepos compared to early-2025 Cursor.

Skills System (Mid-2026)

The Skills primitive — bundled rules, commands, and templates — was formalized in mid-2026. Teams previously cobbled together similar patterns by hand; Skills made it a first-class concept with a directory layout, share patterns, and versioning conventions. Detailed in Chapter 24.

Enterprise Enhancements (Throughout 2026)

A series of features aimed at the Enterprise segment:

- AI-Code-Tracking API for compliance reporting on AI-generated code.
- Enhanced audit logs with SIEM integration patterns.
- Expanded blocklist capabilities (repos, models, MCP servers).

- Self-hosted indexing options for organizations with strict data-locality requirements.
- Custom DPA templates for common compliance regimes.

Privacy Mode Enforcement at Org Level (Q2 2026)

Enterprise admins can enforce Privacy mode org-wide as a policy, removing the per-user override. The feature was heavily requested by regulated-industry customers and shipped in Q2.

Pricing Adjustments (Throughout 2025–2026)

The pricing model has continued to evolve:

- Mid-2025: transition from request quotas to credit pools.
- Late 2025: clarification of per-action credit costs in the dashboard.
- Early 2026: budget alerts and per-action caps.
- Mid-2026: pooled-credit options for Teams; expanded pay-as-you-go.

The credit-pool model has stabilized as the underlying structure; tweaks since have addressed visibility, controls, and team-level features rather than fundamentals.

Anticipated for Late 2026 and Beyond

Speculative; based on public roadmap signals as of book writing:

- Further model picker additions as new frontier models ship.
- Self-hosted Cursor options for the most security-sensitive environments.
- Continued evolution of the agent loop — likely more sophisticated planning, better recovery from off the-rails situations.
- Deeper IDE integrations beyond the current set.
- Expanded Skills ecosystem with public Skill registries.

The roadmap moves fast. Treat the future-looking section as informed speculation, not commitment. Cursor's official changelog at cursor.com/changelog is authoritative for what has actually shipped.

How to Stay Current

The book covers the state as of early 2026. Specific features will keep evolving. To stay current:

1. **Cursor's changelog** at cursor.com/changelog. New features are documented there.
2. **Cursor's blog** at cursor.com/blog. Longer-form announcements and case studies.
3. **The community Discord** (link from cursor.com). Daily discussion of new features and tips.
4. **Cursor's YouTube channel.** Demos and tutorials for new features.
5. **The community subreddit** (r/cursor). Less curated; useful for catching what people are actually struggling with.

The book's recommendation: subscribe to the changelog. The other channels are useful but optional. Most material new-feature changes show up in the changelog within days.

Appendix F: Resources and Further Reading

A curated list of resources for going deeper, staying current, and connecting with the broader Cursor and AI-tooling community.

Official Cursor Resources

cursor.com. The product homepage. Pricing, downloads, basic feature overviews.

cursor.com/docs. The official documentation. Authoritative for current feature behavior; comprehensive but organized feature-by-feature rather than by workflow.

cursor.com/changelog. Release notes for every Cursor update. The single best place to follow what's new.

cursor.com/blog. Longer-form posts about features, customer stories, and product direction.

cursor.com/status. The status page. Check here when something feels broken, to confirm it's not a service issue.

cursor.com/security. Security and privacy posture, sub-processor list, compliance documentation.

cursor.com/pricing. Current pricing for all plans.

Community

Cursor Discord (link from cursor.com). The most active community discussion. Real-time troubleshooting, feature discussion, tips and tricks. Worth joining if you're using Cursor seriously.

r/cursor on Reddit. Less curated than Discord; useful for seeing what people are publicly struggling with. Mixed signal-to-noise.

Cursor on Twitter/X. Official @cursor_ai for announcements; community accounts vary in quality.

Cursor LinkedIn group. Smaller and more enterprise-flavored; useful if you work in larger organizations.

Tutorials and YouTube

Cursor's YouTube channel. Official tutorials and feature demos. Short, focused, kept current.

Independent Cursor tutorial creators. A handful of YouTube creators have built channels around Cursor tutorials. Quality varies; the better ones are workflow-focused rather than feature-tour-focused. Names rotate; search YouTube for "cursor tutorial 2026" and filter for recent uploads with positive signal.

Cursor in production case studies. Various engineering blogs document how teams use Cursor at scale. Search "cursor adoption [company name]" or "cursor case study" for current writeups.

Rules Libraries

github.com/PatrickJS/awesome-cursorrules. A community-curated repository of `.cursorrules` examples (the older format, still useful as a reference). Strong starting point for many stacks.

Other community rules repos. Various individual contributors maintain personal rules libraries on GitHub. Search "cursor rules [language] [framework]" for current options.

MCP Resources

modelcontextprotocol.io. The official protocol documentation. Useful if you're building custom MCP servers.

Cursor's MCP catalog. Browsable from within Cursor (Command Palette → "Install MCP Server"). The vetted starting point.

github.com/modelcontextprotocol/servers. The official repository of reference MCP servers. Source code for many of the standard servers.

Anthropic's MCP documentation at anthropic.com/news/model-context-protocol. Original announcement and conceptual framing.

Books in This Series

For broader context on the AI tooling landscape:

ChatGPT: The Definitive Guide / The Power User Guide — for general LLM use beyond coding.

Claude: The Definitive Guide — for the model behind Cursor's Opus 4.7 option.

Claude Code: The Power User Guide — for the terminal-native alternative to Cursor.

Gemini: The Complete Guide / The Advanced Guide — for Google's frontier model and ecosystem.

Copilot: The Workplace AI Guide / The Advanced Workplace Guide — for the GitHub Copilot alternative.

Perplexity: The Research Playbook / The Advanced Research Playbook — for research-oriented AI use.

NotebookLM: The Complete Guide — for the document-grounded research alternative.

DeepSeek: The Complete Guide — for the cost-quality leader in the picker, plus self-hosting context.

OpenClaw: The Complete Guide — for an alternative AI-powered tooling environment.

AI Systems Playbook — for cross-platform skills and frameworks.

AI Prompt Engineering — for the prompt-craft skill that underlies effective use of any AI tool.

AI for Developers — for the broader category of "developers using AI" beyond Cursor specifically.

AI for Business / AI for Students — for non-developer audiences who interact with the AI ecosystem.

AI Governance for Practitioners — for the policy and compliance dimension.

Securing Agentic AI — for the security dimension of AI agents in production.

All available at press.finnoybu.org.

Adjacent AI-Coding Tools

For comparison or alternative use:

Claude Code at anthropic.com/claude-code. Anthropic's terminal-native agent.

GitHub Copilot at github.com/features/copilot. The Microsoft-aligned alternative.

Windsurf at windsurf.dev (formerly Codeium). The closest direct Cursor competitor.

Cline at github.com/cline/cline. Open-source Cursor-alternative as a VS Code extension.

Continue.dev at continue.dev. Open-source AI assistant for VS Code and JetBrains.

Aider at aider.chat. Open-source CLI agent in the same space as Claude Code.

Void AI at voideditor.com. Newer open-source alternative to Cursor.

Newsletters and Discussion

The Pragmatic Engineer by Gergely Orosz — covers AI tooling adoption alongside other engineering topics.

Stratechery by Ben Thompson — covers the strategic dimension of AI tooling and the platform shifts it represents.

Latent Space by swyx and Alessio — AI engineering newsletter and podcast with significant Cursor coverage.

The Diff and **TLDR AI** — daily AI news roundups.

Academic and Industry Research

For the broader context of AI-assisted development:

arxiv.org searches for "AI code generation," "code completion," "agent-based programming." The research literature is extensive and mostly accessible.

Anthropic, OpenAI, Google DeepMind, DeepSeek research papers. Each lab publishes papers on model capabilities relevant to coding tasks.

MIT, Stanford, CMU research groups working on programming-AI intersections. Various papers on developer productivity, code quality, and human-AI collaboration.

Tools That Pair Well with Cursor

Linear for issue tracking with strong Cursor MCP integration.

Sentry for error monitoring; the MCP integration lets Agent investigate production issues.

Notion for documentation; the MCP integration lets you reference docs without switching tools.

Vercel / Fly.io / Railway for deployment; Background Agents work well for "implement, test, deploy" flows.

Playwright for E2E testing; Agent handles Playwright test creation and debugging well.

TanStack Query (for React projects) — Cursor knows it well and produces strong code.

Drizzle ORM (for TypeScript projects) — query builder Cursor handles confidently.

Prisma (for TypeScript projects) — well-documented, model-friendly schema-first ORM.

FastAPI (for Python projects) — well-typed, model-friendly API framework.

When Resources Conflict

The Cursor ecosystem is fast-moving. Older blog posts, YouTube videos, and Stack Overflow answers may describe behavior that's no longer current. When advice from a community resource conflicts with the official docs, trust the docs. When advice from this book conflicts with anything more recent, trust the more recent source — the book is a snapshot in time.

The exception: workflow advice. Product features change; the underlying disciplines (read every line, scope before invoking, pick the right mode) are stable. When this book and a newer resource agree on workflow but disagree on a feature detail, the workflow advice is probably still right and the feature detail probably needs updating.

Closing Note

The most useful resource is your own usage. After a month or two of deliberate Cursor use with the practices in this book, you'll have intuitions no resource teaches — the feel for which model fits a task, the rhythm of mode switching, the patterns your specific codebase rewards. Trust those intuitions as they develop. The book and the external resources are starting points; your accumulated practice is what turns Cursor into leverage.

Good luck. Ship well.

www.ingramcontent.com/pod-product-compliance
Ingram Content Group UK Ltd.
Pitfield, Milton Keynes, MK11 3LW, UK
UKHW051117290926
15471UKWH00010B/1723

9 798950 533129